MADISON AREA TECHNICAL COLLEGE

50 CMOS IC PROJECTS

MADISON AREA TECHNICAL COLLEGE

50 CMOS IC PROJECTS
by
Delton T. Horn

TAB TAB BOOKS Inc.
Blue Ridge Summit, PA

FIRST EDITION
FIRST PRINTING

Copyright © 1988 by TAB BOOKS Inc.
Printed in the United States of America

Reproduction or publication of the content in any manner, without express permission of the publisher, is prohibited. No liability is assumed with respect to the use of the information herein.

Library of Congress Cataloging in Publication Data

Horn, Delton T.
50 CMOS IC projects / by Delton T. Horn.
p. cm.
Includes index.
ISBN 0-8306-0195-3   ISBN 0-8306-2995-5 (pbk.)
1. Integrated circuits—Amateurs' manuals. 2. Metal oxide semiconductors, Complementary—Amateurs' manuals. I. Title.
II. Title: Fifty CMOS IC projects.
TK9966.H67   1988
621.3815′2—dc19   88-17023
                                    CIP

TAB BOOKS Inc. offers software for
sale. For information and a catalog,
please contact TAB Software Department,
Blue Ridge Summit, PA 17294-0850.

Questions regarding the content of this book
should be addressed to:

Reader Inquiry Branch
TAB BOOKS Inc.
Blue Ridge Summit, PA 17294-0214

Contents

Introduction vii

List of Projects ix

1 Basics of Digital Electronics 1

2 Binary Circuits 23

3 Control Circuits 42

4 Test Equipment 59

5 LED Flashers 76

6 Signal Generator & Music-Making Projects 84

7 Counter Circuits 117

8 Time Keeping Circuits 149

9 Game Circuits   159

10 Miscellaneous Circuits   166

Appendix   185

Index   207

Introduction

Anyone working in electronics these days, professionally or as a hobby, must know something about digital electronics. This book is a collection of 50 practical projects designed around popular CMOS ICs. The operation of each circuit is explained, and in many cases, tips on modifying the project are included.

There is a wide variety of projects in this book, some are quite simple, while others are moderately complex. Often the more complex projects are made up of combinations or variations of the simpler projects. You will find throughout the world of electronics that most complex circuits can be broken down into a collection of simpler circuits.

I strongly encourage experimentation, I hope you use the projects in this book as starting points for your own projects, as well as using them as complete projects on their own. Whenever possible, I have tried to give tips on possible modifications to the basic circuits.

Breadboarding

You could just build the projects directly, constructed on simple perf board with point-to-point wiring, or on simple home-brew printed-circuit boards. However, before you reach for the soldering iron, I strongly suggest that you breadboard each project on a solderless socket first. This is a good idea for any electronics project. It is much easier to trouble-shoot a problem or make modifications to a circuit if you don't waste time de-soldering and re-soldering connections. Besides being time consuming and a real nuisance, this can introduce new problems. Solder bridges become increasingly likely every time you apply the soldering iron, and excessive heat can damage or destroy delicate semiconductor components. It's best to apply heat as infrequently as possible.

Breadboard the circuits first; then, when you are 100 percent sure that everything is right, and the circuit does exactly what you want it to do, solder a permanent version of the project.

Sometimes you will only want to experiment with a circuit for awhile. You may not need a permanent version of the project. In such cases a solderless breadboard is the only rational approach. You can easily reuse the components.

Substituting Components

With very few exceptions (which are *always* indicated in the text) the components used in the projects are not terribly critical. Use a good substitution guide to find equivalent devices for any semiconductors. I have tried to call for only readily available components, but if you have something in your junk box that will do the job just as well, there is no point in running out and buying a brand new part with a slightly different identification number.

All resistors are assumed to be standard 5 percent ¼ watt carbon units. While I haven't tested this, I would imagine 10 percent resistors will work in any of the circuits.

Capacitors are generally assumed to be inexpensive ceramic discs. There is nothing to stop you from using a more expensive, higher quality component, such as a mylar or polystyrene capacitor. I don't think the higher grade capacitors will give you much advantage in the projects presented here. Once again, use what you have handy. In most cases you can substitute values other than those specified in the parts lists.

Large value capacitors (1 μF and up) are standard electrolytic units. For all capacitors, the working voltage should be at least 150 percent of the supply voltage. For example, if you are powering the circuit from a +**9** volt supply, use a capacitor rated for at least 15 volts.

I don't think you should have any trouble with any of the projects presented in this book. I hope you enjoy them and learn from them.

List of Projects

1: Multi-Bit Digital Comparator 26
2: SISO Shift Register 29
3: SIPO Shift Register 31
4: Binary Adder 35
5: Majority Logic Demonstrator 38
6: Touch Switch 43
7: Timed Touch Switch 43
8: Light Activated Gate 46
9: Pulse Delayer 47
10: Automatic Night Light 50
11: Appliance Controller 52
12: Programmable Resistance/Capacitance 53
13: Digital Relay Driver 57
14: Logic Probe 59
15: Simple Frequency Meter 63
16: Digital Frequency Meter 69
17: Capacitance Meter 73
18: LED Flasher 77
19: Pseudo-Random Flasher 81
20: Gated Oscillator 85
21: Tunable Oscillator 86
22: Stepped-Wave Generator 88
23: Triangle-Wave Generator 90

- **24:** Digital Sine Wave Generator 92
- **25:** Toy Organ 96
- **26:** Photo-Theremin 98
- **27:** Four Tone Sequencer 100
- **28:** Ten Step Tone Sequencer 103
- **29:** Random Tone Generator 107
- **30:** Digital Filter 114
- **31:** Decimal Output Counter 119
- **32:** Multi-Digit Decimal Counter 126
- **33:** Decimal Count-Down Timer 129
- **34:** Random Number Generator 132
- **35:** Photoelectric Counter 133
- **36:** Magnetic Reed Switch Counter 139
- **37:** Touch-Switch Counter 141
- **38:** One Hertz Timebase 151
- **39:** One Minute Timer 152
- **40:** Digital Clock 155
- **41:** Electronic Dice 159
- **42:** Roulette Wheel 162
- **43:** Switch Debouncer 166
- **44:** Selectable Delay 167
- **45:** Pulse Width Modulator 171
- **46:** Phase Detector 174
- **47:** Four Step Sequencer 174
- **48:** Frequency Multiplier 177
- **49:** Frequency Divider 177
- **50:** Linear Amplifier 183

50 CMOS IC Projects

1
BASICS
OF DIGITAL ELECTRONICS

Digital electronics is very popular these days, in large part this is due to the so-called "computer revolution." *Digital* suggests high-tech, and the most up-to-date technology.

Analog circuitry certainly isn't obsolete, not by a long shot, it probably never will be. Some functions are best accomplished linearly, others are more suited to digital systems. Some applications can be accomplished by either analog or digital means.

In this book we will concentrate on digital circuits, of course. Some of the applications may seem a little surprising. The crossover line between the analog and digital worlds is becoming increasingly blurred.

Currently, the logic family of choice for most applications is *CMOS* (Complementary Metal-Oxide Semiconductor). CMOS devices are inexpensive, readily available, and easy to work with. What more could a hobbyist want? As the title of this book indicates, all of the projects discussed here will be based on popular CMOS ICs.

WHAT IS A DIGITAL SIGNAL?

Before getting to the actual projects, we should review some basics of digital electronics.

An analog signal such as the one shown in Fig. 1-1, can be comprised of any number of instantaneous values, between 1.0 volt and 2.0 volts, there are an infinite number of possible divisions. For example: 1.11, 1.5, 1.25, 1.72, 1.98, 1.9999, and so on. Between any two voltage points, there is always an intermediate value. Signals usually change smooth and linearly from one value to another, passing through all intermediate values in the process.

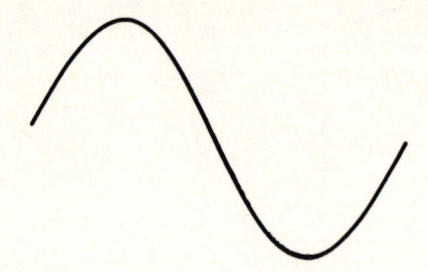

Fig. 1-1. Analog signals can pass through any number of discrete values.

Digital signals are much more limited. A digital signal may only take on one of two possible values. They are called by various names that all mean the same thing (See Table 1-1.):

HIGH	LOW
1	0
+	−
YES	NO

Table 1-1.

Theoretically, digital signals switch instantly from one state to the other, as shown in Fig. 1-2. Real-world devices are not perfect, so there is some finite transition time between states. Except in extreme cases, the transition can be ignored, as it has no particular effect on the action of the digital circuit.

A single digital signal is called a bit. A bit is always either a 1 or a 0. There is never any ambiguity.

A single bit can't carry much information, so a string of bits [either in serial (one after the other) or parallel (all at once)] is often used in digital systems. Each additional bit increases the number of possible bit combinations by a factor of two. Certain bit string-lengths are given special names, demonstrated in Table 1-2.

Sixteen and thirty-two bit "words" are often used in computer systems.

NUMBER BITS	NUMBER OF COMBINATIONS	(NAME)
1	2	bit
2	4	
3	8	
4	16	nibble
5	32	
6	64	
7	28	
8	256	byte

Table 1-2.

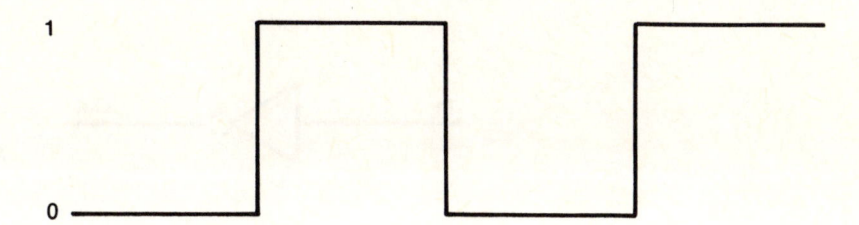

Fig. 1-2. Digital signals use just two discrete values.

LOGIC GATES

The basic unit of digital electronics is the logic gate. A logic gate is a circuit that produces one (or more) digital output whose state is determined by the state(s) of one or more digital input(s). Each possible combination of input states produces an unambiguously predictable output.

Logic gates can be constructed from discrete components, but in modern circuits ICs are almost invariably used. They are smaller, cheaper, and more reliable, so why go to the extra bother of wiring up discrete component gates?

The simplest logic gate is the buffer, which is shown in Fig. 1-3. This device has one input and one output. The output state is always the same as the input state. If the input is LOW, the output will be LOW. If the input is HIGH, the output will be HIGH. This certainly sounds like a pointless exercise. After all, if the logic state is completely unaffected by the gate, then why bother with the gate at all? Actually, the buffer gate is used similarly to the analog buffer amplifier or isolation amplifier. The buffer gate is used to increase the *fan-out* of a digital output. Any digital output can drive just so many inputs before it becomes loaded down. If an output has a fan-out of 10, it can drive 10 buffers, each of which can drive 10 inputs. This gives a total effective fan-out of 100.

The buffer gate is not very widely used in practical digital circuits, but it does show up occasionally.

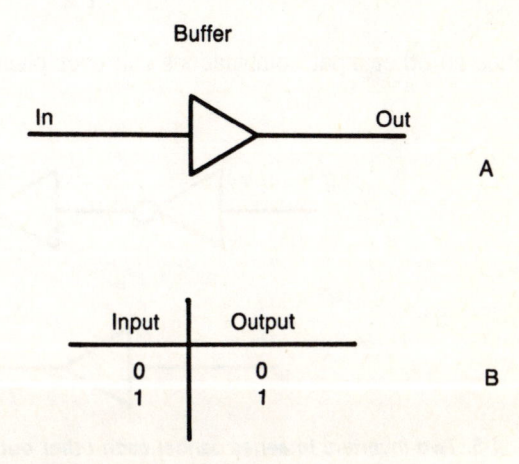

Fig. 1-3. The simplest digital gate is the buffer.

Inverter

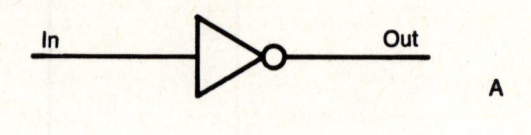

A

Input	Output
0	1
1	0

B

Fig. 1-4. Inverter output is the opposite of the input.

There is another one-input/one-output logic gate, that is widely used. This is the inverter, shown in Fig. 1-4. As the name suggests, this device inverts the logic state. The output always has the opposite state of the input. If the input is LOW, the output is HIGH. If the input is HIGH, the output is LOW. Inverters are used in a great many digital circuits.

Two inverters in series cancel each other out. The net result is the same as a buffer, shown in Fig. 1-5.

Most basic logic gates have two inputs and one output. With two gates, there are four possible input combinations:

```
0 0
0 1
1 0
1 1
```

Notice no other input combinations can ever occur.

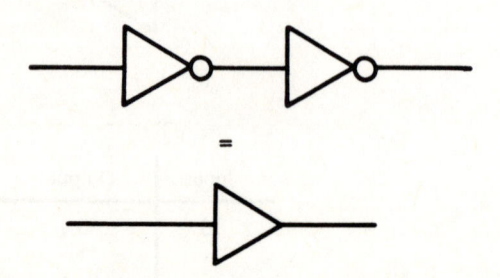

Fig. 1-5. Two inverters in series cancel each other out and act as a buffer.

4

A chart showing the output for each possible input combination is called a *truth table*.

There are several standard input/output patterns that are widely used.

Figure 1-6 shows the AND gate. The name is somewhat self-explanatory. The output is HIGH if and only if input A *AND* input B are HIGH. If either (or both) of the inputs is LOW, the output will be LOW.

AND gates can have more than two inputs. Figure 1-7 shows a three-input AND gate. It works in the same manner as the two-input version. The output is HIGH if and only if *all* of the inputs are HIGH. If any one (or more) of the inputs are LOW, the output will be LOW.

A three-input AND gate can be built up from a pair of two-input AND gates, as illustrated in Fig. 1-7C.

A popular combination of gates is shown in Fig. 1-8. Inverting the output of an AND gate gives us the opposite pattern of an unmodified AND gate. Its output is HIGH *unless* input A *and* input B are both HIGH. As long as at least one input is LOW, the output will be HIGH. We could call this a NOT AND operation. It is such a widely used function that the standard name is contracted to NAND.

Dedicated NAND gate ICs are very widely available. In fact, this is probably the most popular type of logic gate. The symbol for an NAND gate is shown in Fig. 1-8B.

Another basic type of multiple-input gate is the OR gate, which is shown in Fig. 1-9. Once again, the name suggests the operation. The output is HIGH if either *OR* both of the inputs are HIGH. The output is LOW if, and only if, both inputs are LOW. Some OR gates have more than two inputs. Figure 1-10 shows a three-input OR gate. Operation is the same as in the two-input OR gate. The output is HIGH whenever at least one of the inputs is HIGH. The output is LOW only when *all* of the inputs are LOW. A three-input OR gate can be created from a pair of two-input gates. Inverting the output of an OR gate gives us a NOR gate. This is illustrated in Fig. 1-11. The output is HIGH if, and only if, neither input A *NOR* input B is HIGH. If either (or both) of the inputs is HIGH, the output will be LOW.

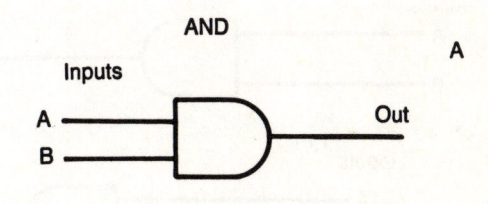

Inputs A B	Output
0 0	0
0 1	0
1 0	0
1 1	1

Fig. 1-6. A popular multiple-input gate is the AND gate.

Three-input AND

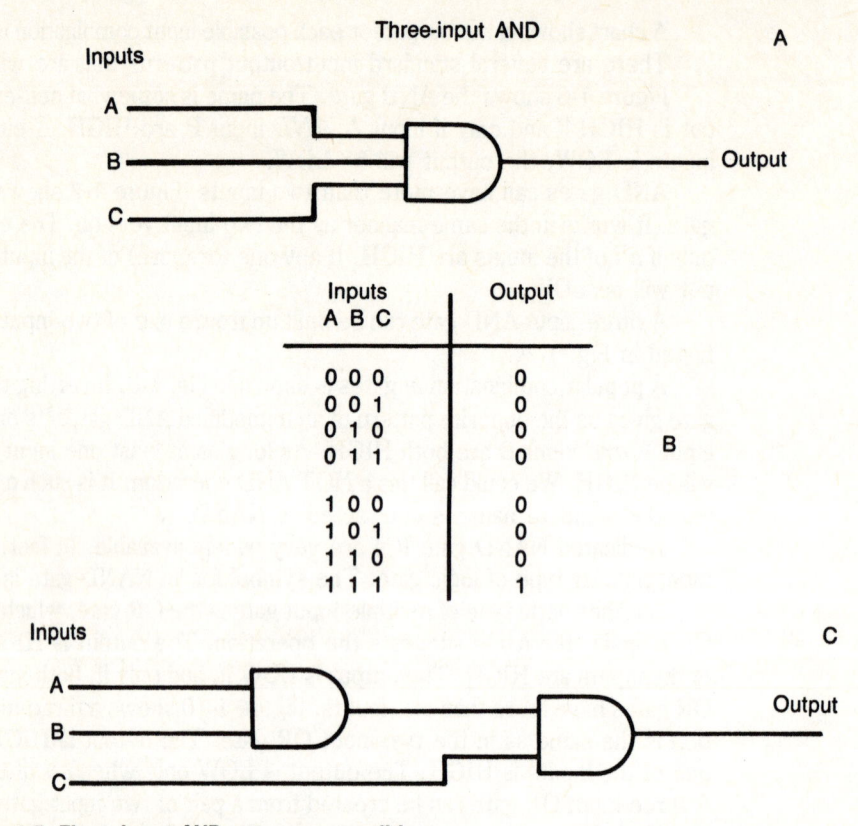

Inputs A B C	Output
0 0 0	0
0 0 1	0
0 1 0	0
0 1 1	0
1 0 0	0
1 0 1	0
1 1 0	0
1 1 1	1

Fig. 1-7. Three-input AND gates are possible.

NAND

Inputs A B	Output
0 0	1
0 1	1
1 0	1
1 1	0

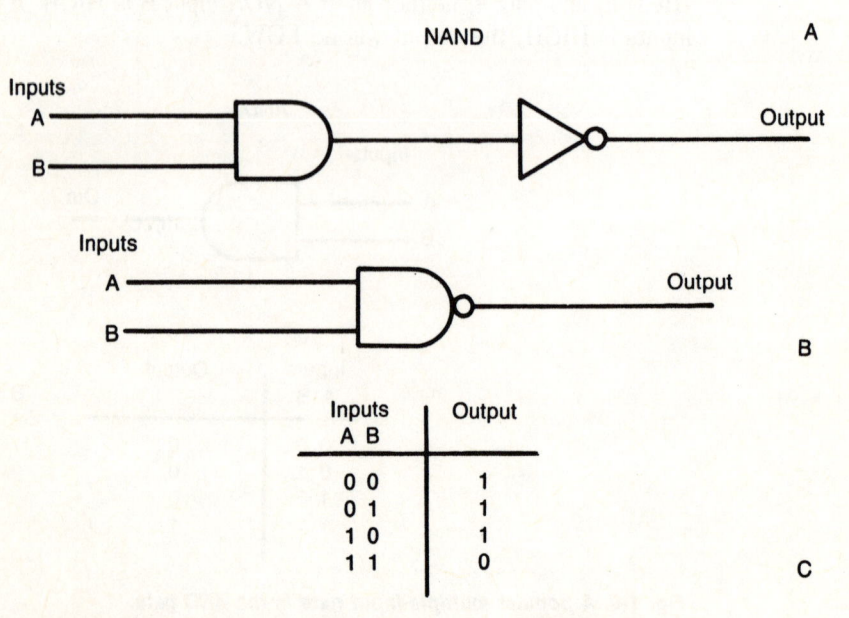

Fig. 1-8. Inverting the output of an AND gate produces a NAND gate.

6

OR

Inputs
A ————
B ————
              Output

Inputs A B	Output
0 0	0
0 1	1
1 0	1
1 1	1

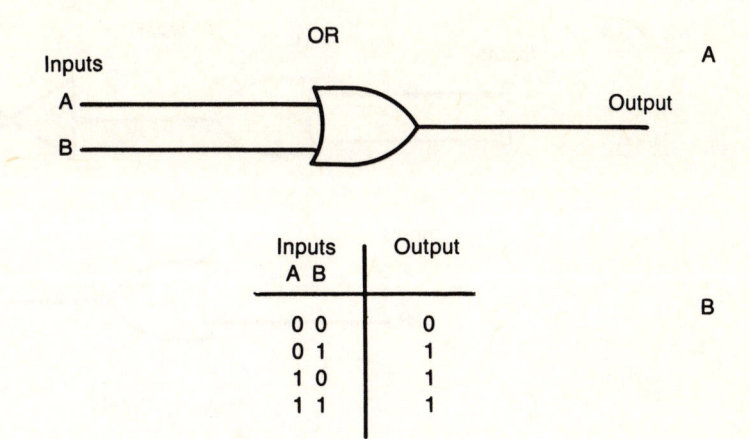

Fig. 1-9. Another popular multiple-input gate is the OR gate.

Three-Input OR

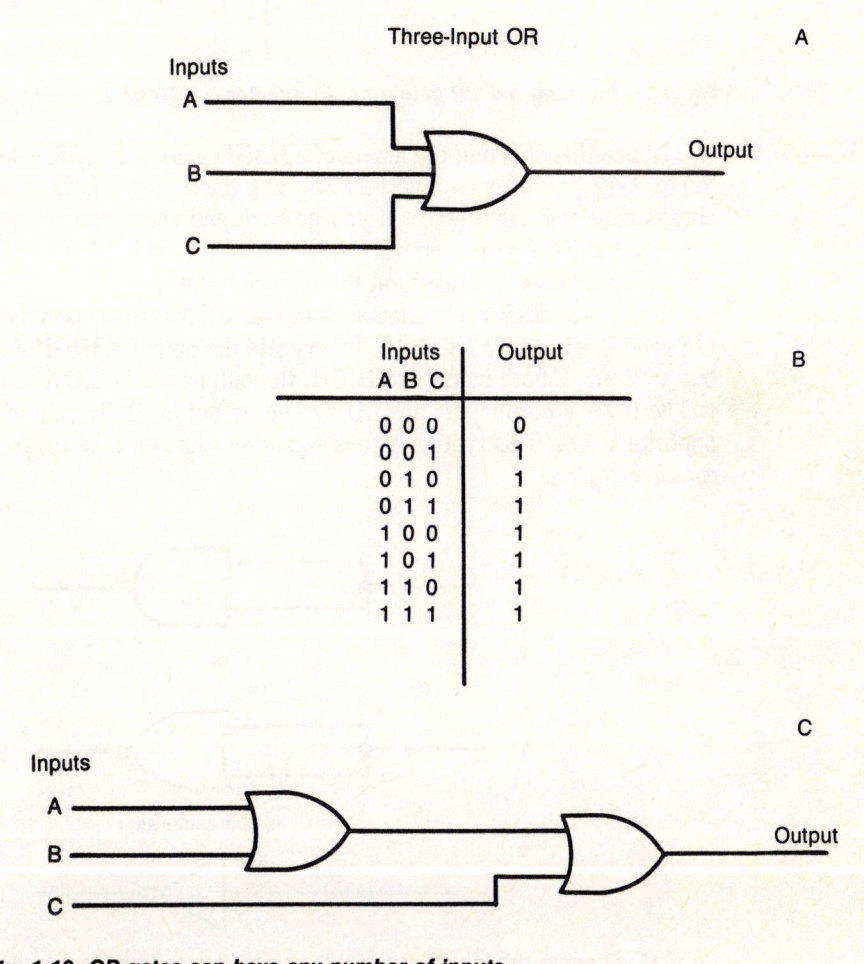

Inputs
A ————
B ————
C ————
              Output

Inputs A B C	Output
0 0 0	0
0 0 1	1
0 1 0	1
0 1 1	1
1 0 0	1
1 0 1	1
1 1 0	1
1 1 1	1

Inputs
A ————
B ————
C ————
              Output

Fig. 1-10. OR gates can have any number of inputs.

NOR

A

B

C

Inputs A B	Output
0 0	1
0 1	0
1 0	0
1 1	0

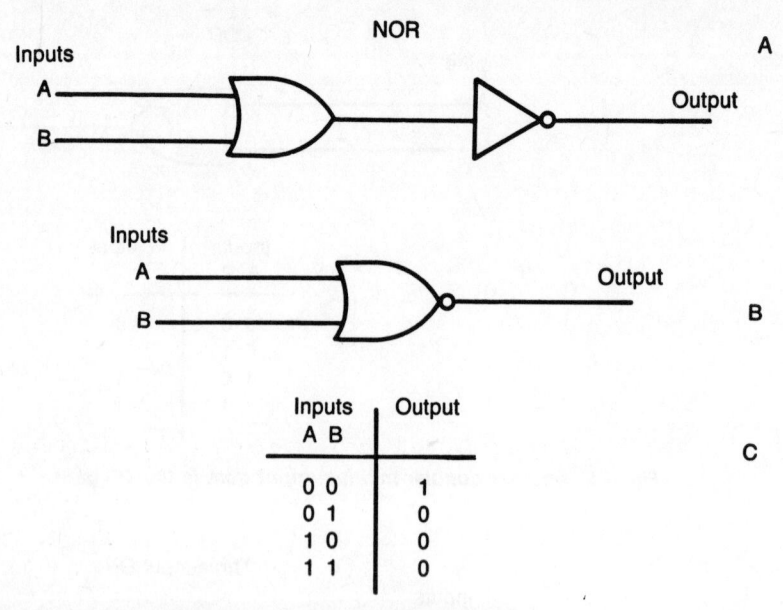

Fig. 1-11. Inverting the OR gate output produces a NOR gate.

Notice that shorting the inputs of a NAND gate or a NOR gate together, as shown in Fig. 1-12 produces an inverter. Shorting the inputs together ensures that they will always have the same state. If you go back and check the truth tables of the NAND gate and the NOR gate, you will see that in either case, when the two inputs are identical, the output always takes on the opposite state.

There is a specialized variation on the basic OR gate. This is the X-OR, or *eXclusive-OR* gate, illustrated in Fig. 1-13. In this gate the output is HIGH if either input is HIGH, but *not both*. If both inputs are HIGH, the output will be LOW. In addition, the output will be LOW if both inputs are LOW. The output is HIGH only when the inputs are at opposite states. This type of gate is sometimes known as a *difference detector*, or *nonequality gate*.

or

is the same as

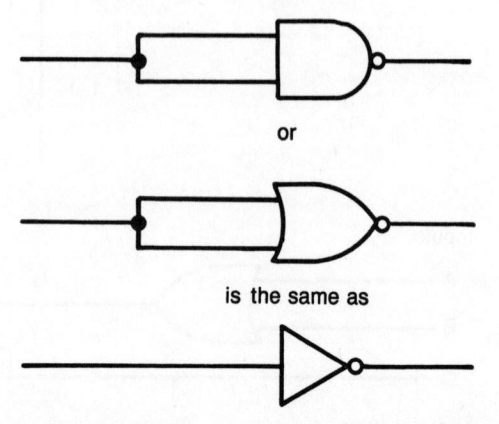

Fig. 1-12. Shorting the inputs of a NAND or a NOR gate together simulates an inverter.

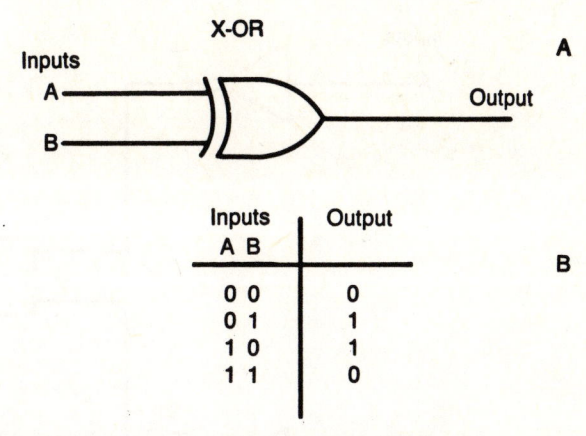

Fig. 1-13. A specialized variation on the OR gate is the X-OR gate.

The basic logic gates can be combined to come up with non-standard input/output patterns. Any truth table imaginable can be achieved with some combination of basic logic gates. Generally, there is more than one way to achieve any truth table pattern.

A simple non-standard logic gate is illustrated in Fig. 1-14. One of the two inputs is inverted before it is fed into an AND gate. The operation of this circuit could be described as A AND NOT-B. The output is HIGH if, and only if, input A is HIGH *and* input B is LOW. All other input combinations will result in a LOW output.

If we invert both of the inputs to an AND gate, as shown in Fig. 1-15, we get an interesting result. The operation could be described as NOT-A AND NOT-B. The output is HIGH if, and only if, input A is NOT HIGH (LOW) AND input B is NOT HIGH (LOW). In other words, the output is HIGH only when neither A *nor* B is HIGH. An AND gate with both inputs inverted functions as a NOR gate. Similarly, if we invert both

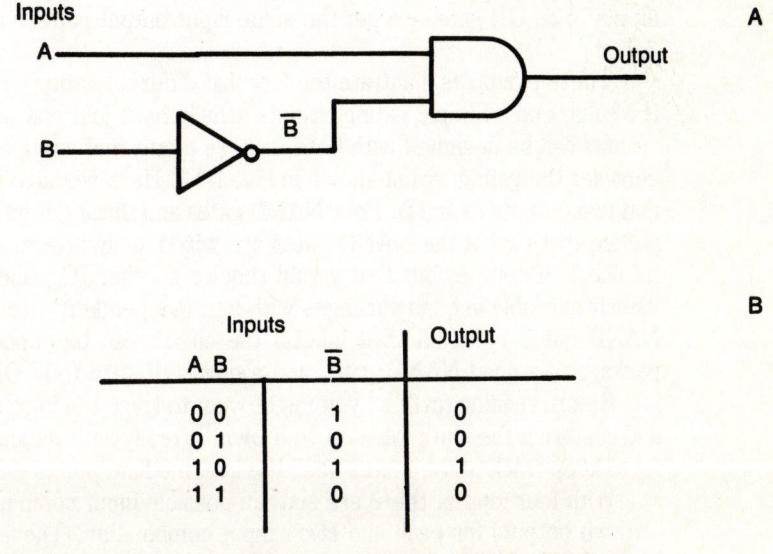

Fig. 1-14. Basic gates can be combined to create non-standard patterns.

9

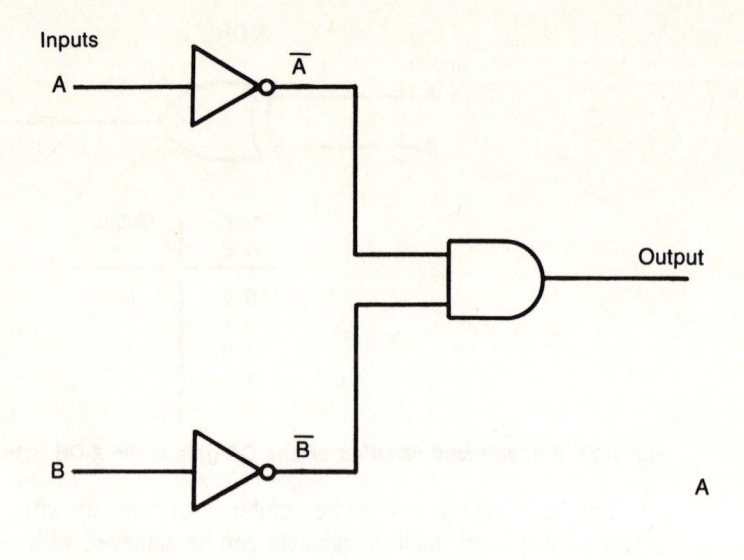

Inputs		Inputs		Output
A	B	$\bar{A}$	$\bar{B}$	
0	0	1	1	1
0	1	1	0	0
1	0	0	1	0
1	1	0	0	0

Fig. 1-15. Inverting both inputs to an AND gate produces a NOR gate.

inputs to an OR gate, we get the same input/output pattern as a NAND gate. See Fig. 1-16.

These examples illustrate the fact that different gating circuits can be used to reach the same ends. Simple gating circuits usually have just one output. But complex gating circuits can be designed with both multiple inputs and multiple outputs. As an example, consider the gating circuit shown in Fig. 1-17. Here we have four inputs (A through D), and two outputs (e and f). Four NAND gates and three OR gates are used in this circuit. Notice that two of the NAND gates are wired as inverters. Dedicated inverters could be used, of course, but that would require another IC package. Two-input gates are usually available in quad packages with four independent gates in a single chip. By using NAND gates (with shorted inputs) the circuit can be constructed from just two IC packages—a quad NAND gate, and a quad OR gate (one OR gate is unused).

Before reading further, you might want to trace the logic signals through the circuit and construct the truth table on your own. It really isn't difficult, as long as you're careful to keep track of the states at all the intermediate points (gate outputs) in the circuit.

With four inputs, there are sixteen possible input combinations. Remember, there are two outputs for each and *every* input combination. These outputs may or may not be at the same state at any given time.

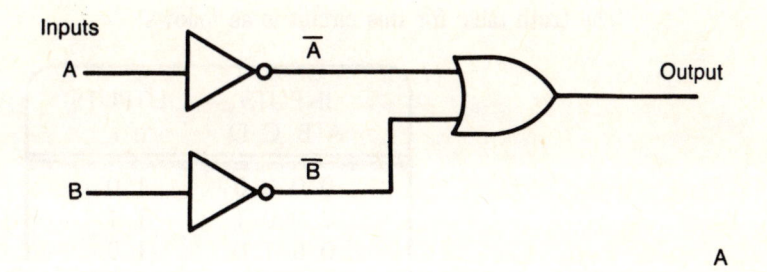

Fig. 1-16. Inverting both inputs of an OR gate produces a NAND gate.

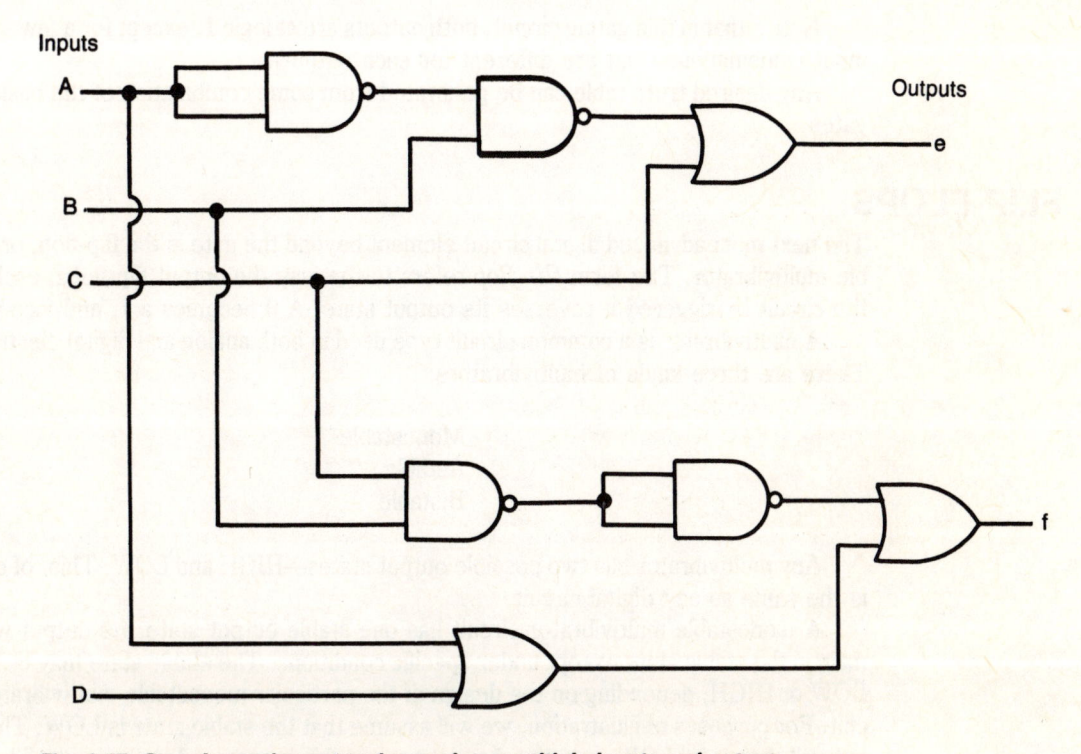

Inputs		Output
A B	$\overline{A}$ $\overline{B}$	
0 0	1 1	1
0 1	1 0	1
1 0	0 1	1
1 1	0 0	0

Fig. 1-17. Complex gating networks can have multiple inputs and outputs.

11

The truth table for this circuit is as follows:

INPUTS A B C D	OUTPUTS e f
0 0 0 0	1 0
0 0 0 1	1 1
0 0 1 0	1 0
0 0 1 1	1 1
0 1 0 0	0 0
0 1 0 1	0 1
0 1 1 0	1 1
0 1 1 1	1 1
1 0 0 0	1 1
1 0 0 1	1 1
1 0 1 0	1 1
1 0 1 1	1 1
1 1 0 0	1 1
1 1 0 1	1 1
1 1 1 0	1 1
1 1 1 1	1 1

Table 1-3.

Notice that in this gating circuit, both outputs are at logic 1, except for a few specific input combinations (that are different for each output).

Any desired truth table can be generated from some combination of the basic logic gates.

FLIP-FLOPS

The next most advanced digital circuit element beyond the gate is the flip-flop, or bistable multivibrator. The term *flip-flop* refers to the way the output flip-flops, each time the circuit is triggered it reverses its output state. A 0 becomes a 1, and vice versa.

A multivibrator is a common circuit type used in both analog and digital electronics. There are three kinds of multivibrators:

> Monostable
> Astable
> Bistable

Any multivibrator has two possible output states—HIGH and LOW. This, of course is the same as any digital circuit.

A monostable multivibrator circuit has one stable output state. Its output will remain in the stable state except under specific conditions. The stable state may be either LOW or HIGH, depending on the design of the particular monostable multivibrator circuit. For purposes of illustration, we will assume that the stable state is LOW. The output will always be LOW, except when a trigger pulse is received at the input. Then the

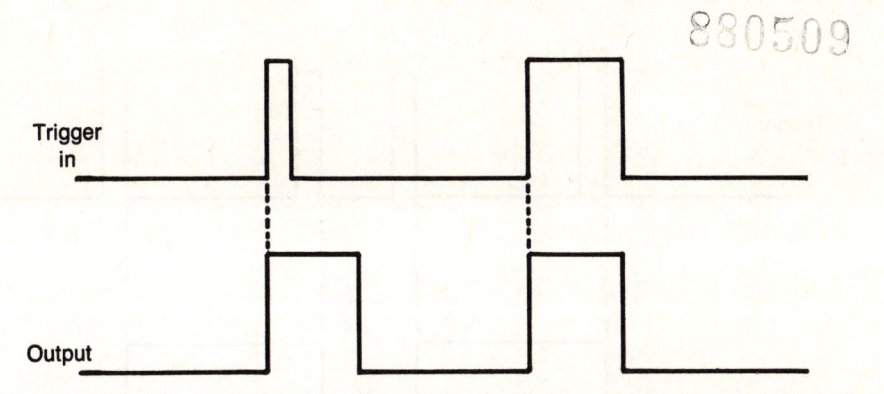

Fig. 1-18. A monostable multivibrator generates an output pulse only when it is externally triggered.

output will go HIGH for a specific period of time determined by component values within the multivibrator circuit. The output pulse will always have the same length, regardless of the length of the input trigger pulse. The output of a typical monostable multivibrator is illustrated in Fig. 1-18. A monostable multivibrator is a basic timer circuit.

An *astable multivibrator* has no stable output states. It can't hold the output either LOW or HIGH very long. The output keeps switching back and forth between states as long as power is applied to the circuit. The rate of this switching is determined by component values within the astable multivibrator circuit. The operation of an astable multivibrator is illustrated in Fig. 1-19. Notice that there is no input trigger pulse for an astable multivibrator circuit.

By now you should have guessed that a bistable multivibrator circuit has two stable output states. The output can be held either LOW or HIGH indefinitely (as long as power is applied to the circuit, of course). The output changes state only when an input trigger pulse in received by the circuit. Each time a bistable multivibrator is triggered, its output changes state. The output of a typical bistable multivibrator is illustrated in Fig. 1-20.

Unlike the other types of multivibrators, a bistable multivibrator has no internal time constant. Its timing is determined solely by the input trigger pulses.

In a sense, a bistable multivibrator, or flip-flop can be considered a simple one-bit memory. As long as the power supply is not interrupted, the circuit will "*remember*" its previous state until it is triggered into a new state. Some computer memories use circuitry quite similar to flip-flops.

Flip-flops can be built up from basic gates. In fact, *all* digital circuits can be constructed from individual gates. Many dedicated flip-flop ICs are available, but their internal circuit is still a collection of basic digital gates. For our discussion, we will assume the flip-flop circuits under discussion are built up from discrete digital gates, so you will better understand how the circuitry works.

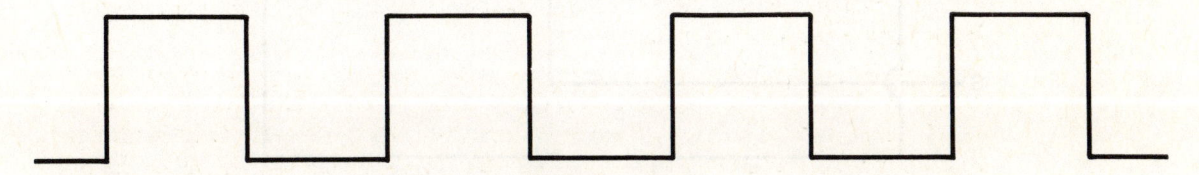

Fig. 1-19. An astable multivibrator automatically switches back and forth between output states.

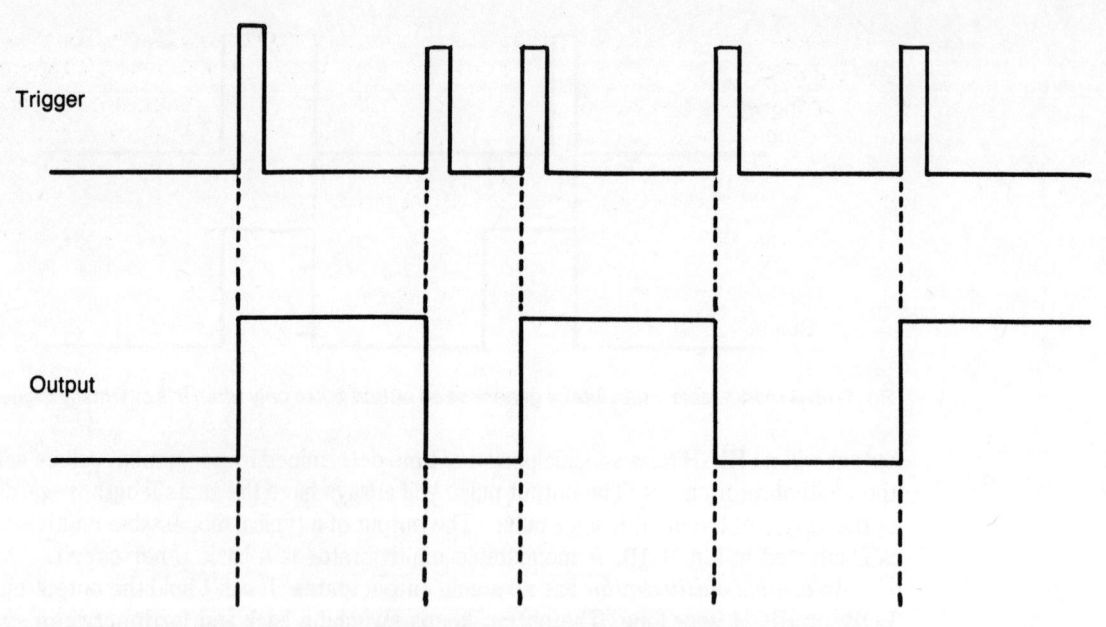

Fig. 1-20. A bistable multivibrator reverses its output state each time it is triggered.

There are many different types of flip-flop circuits. The simplest is the one shown in Fig. 1-21. This circuit is known as a *R-S flip-flop*. It has two inputs—S (Set) and R (Reset).

Notice that this circuit also has two outputs. The main output is labelled "Q." The other output is simply an inversion of the main output. It is labelled "$\overline{Q}$," or "Q-NOT." The $\overline{Q}$ output will always have the opposite state as the Q output. These two outputs are standard for all types of flip-flops.

Essentially, the S (Set) input sets the Q output to 1 ($\overline{Q}$ = 0), and the R (Reset) input resets the Q output back to 0 ($\overline{Q}$ = 1). A HIGH signal on the appropriate input forces the output to change.

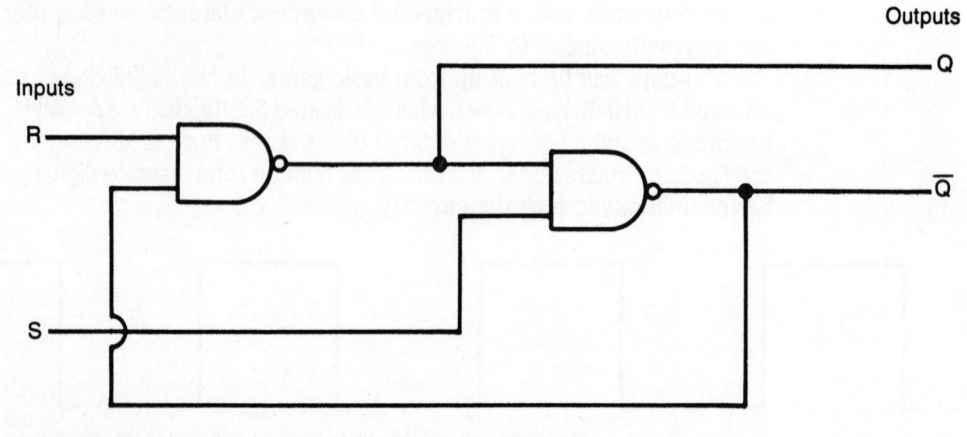

Fig. 1-21. The most basic type of flip-flop is the RS type.

This leaves two other possibilities, what if the same logic signal simultaneously appears on both inputs? If both the S and the R inputs are HIGH, nothing at all will happen. The outputs will not change state. Now, can you guess what will happen if both inputs are LOW? This combination of inputs will "confuse" the circuit, and the output states will be unpredictable. Obviously this is extremely undesirable. This input combination must be avoided at all times. It is called a *disallowed state*.

We can summarize the operation of the R-S flip-flop with the truth table, Table 1-4:

INPUTS R S	OUTPUTS Q $\overline{Q}$	
0 0	? ?	DISALLOWED STATE
0 1	1 0	
1 0	0 1	
1 1	NO CHANGE	

Table 1-4.

The R-S flip-flop circuit we have been looking at is made up of a pair of NAND gates. It is also possible to construct an R-S flip-flop from two NOR gates, as shown in Fig. 1-22.

This circuit functions in basically the same way as the previous one. The only practical difference is that the *disallowed* state and the *no change* state are reversed, as the truth table of Table 1-5 illustrates.

Using an R-S flip-flop tends to be rather problematic. Avoiding the disallowed states can often be difficult. Transitions in the input signals can also create problems. If the two inputs do not change in perfect synchronization an undesired pulse might appear on the output lines.

One solution would be to add an *ENABLE* input. The signals on the R and S inputs will be recognized only when the circuit is enabled. In effect we now have a gated R-S flip-flop. A circuit of this type is shown in Fig. 1-23.

The truth table for this circuit is identical to the original R-S flip-flop (NAND gate version), except the inputs will be recognized only when the ENABLE input is HIGH. If the ENABLE input is LOW, the signals on the R and S input lines will simply be ignored.

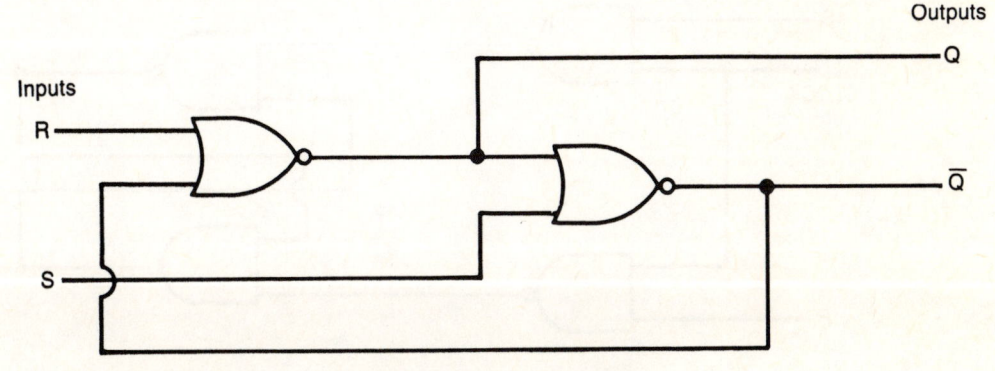

Fig. 1-22. An RS flip-flop can be constructed from NOR gates.

INPUTS R S	OUTPUTS Q Q̄	
0 0	NO CHANGE	
0 1	1 0	
1 0	0 1	
1 1	? ?	DISALLOWED STATE

Table 1-5.

Flip-flops in general, and especially R-S type flip-flops are often called *latches*.

Another popular type of flip-flop circuit is the D-type flip-flop illustrated in Fig. 1-24.

This type of flip-flop has just a single main input, labelled D (for Data). A second input is almost always included to ENABLE the circuit. The signal at the D input is recognized only when the ENABLE input is HIGH. The ENABLE input is often labelled *CLOCK*. When this input is LOW, the D input has no effect on the outputs.

Assuming that the flip-flop is enabled, the Q output will always have the same logic state as the D input. Of course, the Q̄ output will always have the opposite state.

Suppose we feedback the Q̄ signal to the D input and use a string of pulses as the clock input, as shown in Fig. 1-25, the pin numbers given here are for a CD4013 dual D-type flip-flop IC. Two independent D-type flip-flops are contained on a single chip.

Assume that initially Q = 1, so Q̄ = 0. When the clock pulse goes HIGH, the circuit will look at the signal on the D input. This signal is LOW, so the Q output goes LOW, and Q̄ goes HIGH. On the next clock pulse, the HIGH signal at the D input forces the Q output HIGH and the Q̄ output goes LOW again. This process will repeat for as long as new pulses appear at the CLOCK input. The input and output signals are illustrated in Fig. 1-26.

For every two input (CLOCK) pulses, a single pulse appears at the output. The output signal will have half the frequency of the CLOCK signal. The frequency has been effectively divided by two.

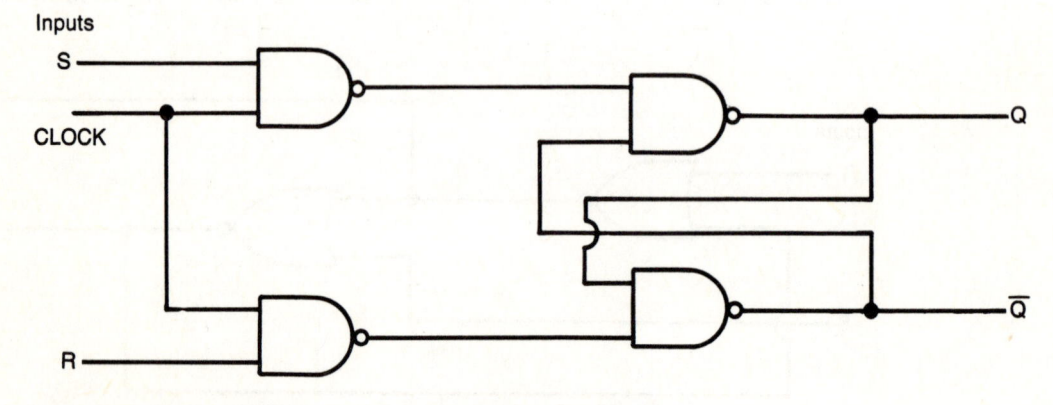

Fig. 1-23. A better flip-flop can be made by adding a CLOCK input.

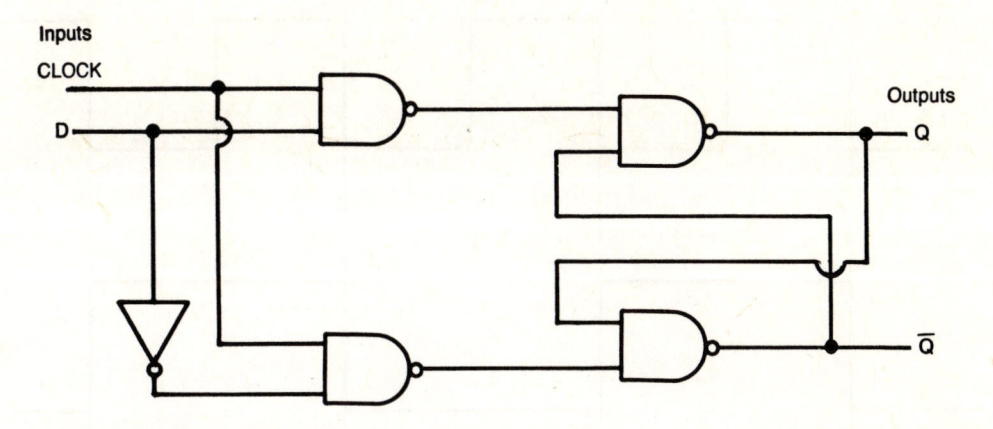

Fig. 1-24. Another common flip-flop is the D-type.

There are several other types of flip-flop circuits, we won't go into here. You will probably encounter J-K flip-flops from time to time. This is essentially an extension of the basic R-S flip-flop. In addition to the R and S inputs, there are additional inputs labelled J and K. These inputs are sometimes called *preset* and *clear*. Curiously, no one seems to know just what "J" and "K" stand for, but those are the well-established names for these inputs, so we're more or less stuck with them.

COUNTERS

In discussing the D-type flip-flop, we learned that it can be used to divide the clock frequency by two. Suppose we chain two of these circuits in sequence, as illustrated in Fig. 1-27.

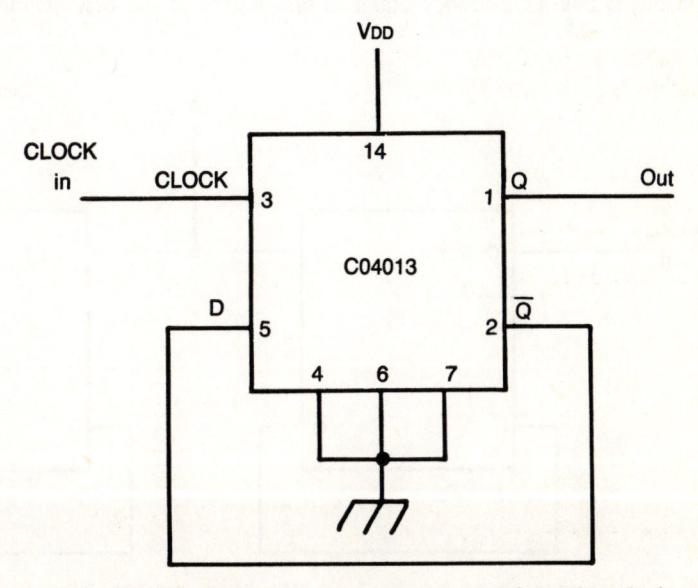

Fig. 1-25. This circuit reverses its output state on each incoming clock pulse.

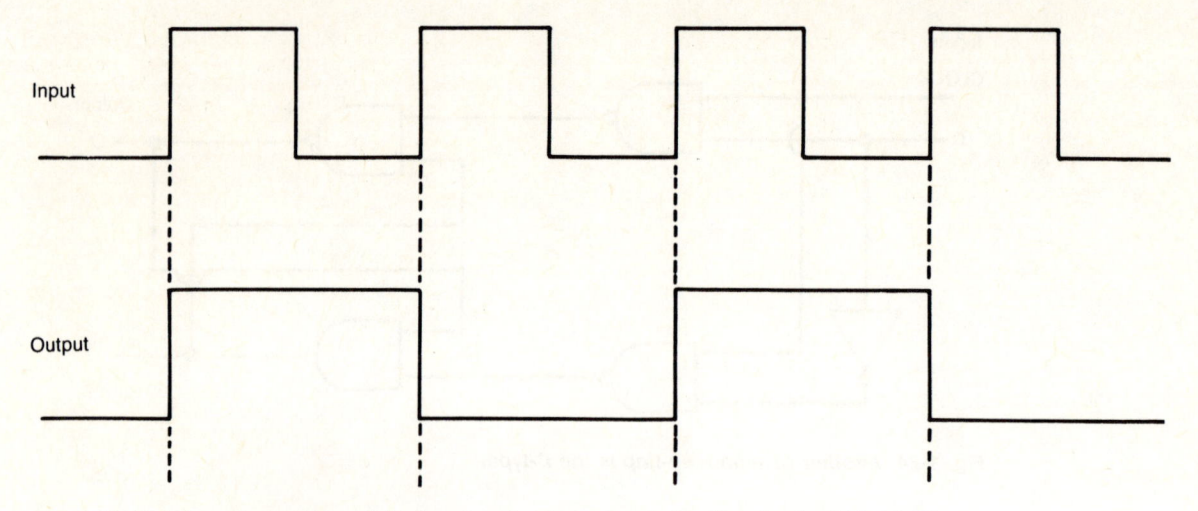

Fig. 1-26. This is a comparison of the input and output signals for the circuit of Fig. 1-25.

Flip-flop A will divide the clock frequency by two. This signal will appear at the output labelled "A," and will also be fed into the input of flip-flop B. Flip-flop B divides signal A by two. Output B will be half of half of the original clock frequency. That is:

$$\begin{align} A &= \text{CLOCK} \div 2 \\ B &= A \div 2 \\ &= (\text{CLOCK} \div 2) \div 2 \\ &= \text{CLOCK} \div 4 \end{align}$$

Output B has a frequency equal to one fourth of the original clock frequency.

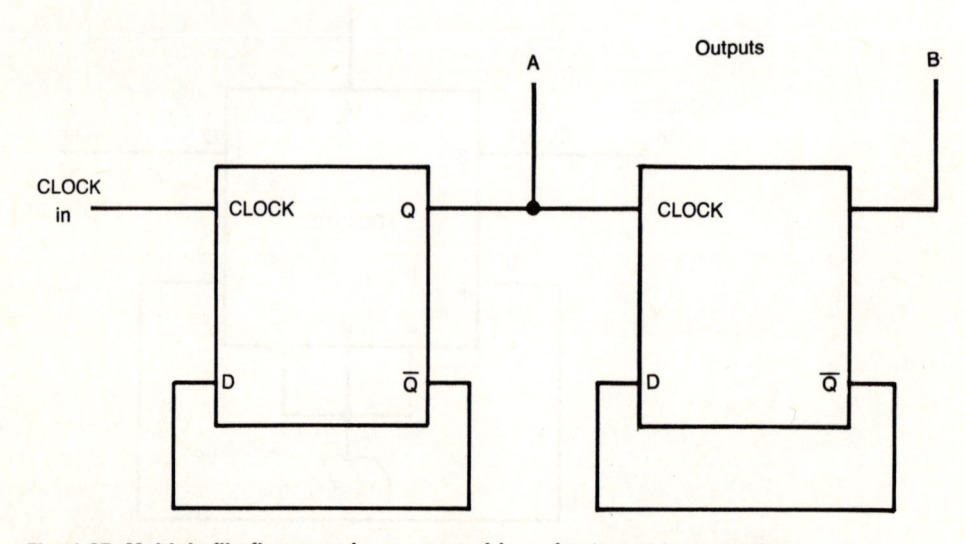

Fig. 1-27. Multiple flip-flops can be connected in series to create a counter.

While potentially useful in certain applications, this certainly isn't very exciting, or unexpected. But let's take a look at what's happening from another angle. The following is a summary of what happens on each successive clock pulse in Table 1-6:

CLOCK PULSE #	OUTPUTS B A
0	0 0
1	0 1
2	1 0
3	1 1
4	0 0
5	0 1
6	1 0
7	1 1
8	0 0

Table 1-6.

and so on.

The outputs count the input (CLOCK) pulses in binary form. Because this is such a small counter it can't count past three (11). When the maximum count is exceeded, the counter resets itself (all outputs to 0), and starts over.

This is the basic principle behind all binary counter circuits. The counter can be extended as much as you like, simply by adding additional flip-flop stages. Each additional stage increases the maximum count by a factor of two. The maximum count, or modulo of a counter is the number of counting steps, including zero. In our example above, the modulo is four, because there are four counting steps (00, 01, 10, and 11). Here are some additional examples in Table 1-7:

NUMBER OF STAGES	MODULO
1	2
2	4
3	8
4	16
5	32
6	64

Table 1-7.

and so on.

If you happen to need a counter that has a modulo that is not an even power of two, some external gating circuitry can be added to forcibly reset the counter (all outputs to 0) after any specified count. Obviously, the modulo of the counter without the reset gating network must be equal to the next highest power of two. For example, let's say we need a counter with a modulo of 12. We would start out with a four stage counter (modulo = 16), then add the necessary gates to force the counter to rest to 0000 after a count of 1011 (decimal 11).

We have only discussed binary counters. Other types of counters are available, including decimal counters, and BCD (Binary-Coded-Decimal) counters. Many different types of counters are available in IC form.

WHY CMOS?

CMOS is one of several *logic families*. A logic family refers to the type of circuitry used to form the gate. Two early logic families were RTL and DTL. RTL stands for *Resistor-Transistor-Logic*. Each gate is made up of a resistor and a transistor. Similarly, DTL is *Diode-Transistor-Logic*, using gates made up of a diode and a transistor.

RTL and DTL were never very popular. They suffered from low density (only relatively simple circuits could be integrated on a chip) and fairly low switching speeds.

Digital electronics didn't really take off until the development of the TTL logic family. TTL stands for *Transistor-Transistor-Logic*. Each gate is made up of two transistors (actually a single transistor with dual emitters is used). TTL offers fair chip density, and high switching speeds. Special variants on TTL are designed for even higher switching speeds.

TTL is not an unmixed blessing, its power supply requirements are very stringent. TTL ICs can be powered only by a tightly regulated +5 volt supply. If the supply voltage even momentarily exceeds +5.25 volts, or drops below +4.5 volts the chip can be damaged. In addition, TTL gates tend to consume a fairly high amount of current.

TTL circuitry does not really permit very high density chips. Complex ICs, such as microprocessors, aren't very practical using TTL technology.

Recently, CMOS has overtaken TTL as the logic family of choice. The name CMOS stands for *Complementary-Metal-Oxide-Semiconductor*. It really isn't very important for the hobbyist to understand what this technical term means. It simply refers to a specific way of etching individual semiconductors on the silicon chip of an integrated circuit.

CMOS chips can be packed with very high density, allowing a great many functions on a single IC. The major advantages of the CMOS logic family lie in the power supply requirements. While a well-regulated power supply is advisable for any digital circuit, CMOS devices will accept a wide range of supply voltages. Anything in the range of +3 volts to +18 volts can be used. For best results, it is a good idea to use a supply voltage somewhere in the middle of the acceptable range.

An even more important advantage of CMOS devices is that they consume very little power. A typical CMOS gate consumes only about **0.1 mA**, considerably less than a comparable TTL gate.

No technology is perfect, there are always trade-offs. CMOS gates have lower maximum switching speeds than comparable TTL gates. Fortunately modern CMOS devices can switch at rates high enough for the majority of practical applications.

SPECIAL PRECAUTIONS

CMOS devices require certain special handling procedures. These were more critical with early devices. Newer CMOS chips include a lot more internal protection, but special handling is still strongly advised. It's always better to be safe than sorry.

CMOS devices are very sensitive to static electricity, newer CMOS ICs are not quite as sensitive, but they can still be damaged by a burst of static electricity. CMOS chips should always be stored with their pins shorted together. Use conductive foam or a conductive holder of some kind. *Never* store CMOS ICs in a non-conductive plastic tray or bag. Do not use non-conductive foam.

When actually working on a circuit, you will occasionally need to set a chip down, place CMOS ICs pin down on a sheet of aluminum foil or on a conductive tray. Avoid touching the pins, especially if you are not using a grounding strap. Use a battery-powered or grounded soldering iron. Better yet, use sockets, and avoid soldering directly to the pins of a CMOS IC.

You should never install any IC into a circuit while power is applied. This is particularly important for CMOS ICs.

Always ground *all* unused inputs on any CMOS chip. Alternatively, unused inputs can be connected to V_{DD} (the positive supply voltage). No CMOS input should *ever* be left floating. Operation of the circuit will be erratic. Stray signals will almost certainly be picked up by floating inputs. Besides causing erratic operation, such stray signal pickup will increase power consumption.

INPUT SIGNAL RESTRICTIONS

Any input to a CMOS device must be within the limits of the power supply, that is the input signal must never drop below ground (0 volts) or rise above V_{DD} (+V). No signal should ever be applied to the input of a CMOS device when no power is being supplied to the chip. Even a momentary input signal to an unpowered chip could cause irreparable damage.

There is a practical limit to the switching speed of any digital gate. The frequency of input signal must never exceed the maximum operating frequency of the CMOS device. The maximum operating frequency is dependent on the actual supply voltage. A typical CMOS device has a maximum output frequency of about 1 MHz when the supply voltage is +5 volts. Increasing the supply voltage to +15 volts raises the maximum operating frequency of the device to about 5 MHz.

Input signals with fast rise and fall times are greatly preferred. Input signals that change state slowly will increase the amount of power consumed by the gate. As a rule of thumb, the rise and fall times for a signal to a CMOS device should be 15 microseconds, or better.

POWER SUPPLIES

Because of the low power consumption, many CMOS circuits can practically be powered from batteries. When only a few gates are used in the circuit, long battery life can be expected. Remember, if a lot of CMOS gates are included in the circuit, overall power consumption will be increased. In addition, off chip devices such as LEDs, and relays can increase the circuit's power consumption by a great deal. An ac-to-dc power supply can be used. Some voltage regulation is advised. Voltage regulator ICs are readily available at reasonably low cost.

Whatever the power source, bypass capacitors are a good idea to prevent problems from surges, or glitches from rapid switching. A 1 μF capacitor should be placed across the power supply's output. For maximum protection, a 0.1 μF capacitor should be placed between the V_{DD} and ground pins of each individual CMOS IC in the circuit. This bypass capacitor should be mounted physically as close to the body of the IC as possible.

In some circuits (especially relatively simple, low-frequency circuits) individual bypass capacitors for each and every IC can be overkill. Alternatively, you can place a single 0.1 μF capacitor in parallel with the 1 μF capacitor across the power supply output.

Note that the bypass capacitors are *not* shown in the schematic diagrams of the projects in this book. In some cases, the power connections to the chips are not shown (this is done to simplify the diagrams). Since these connections are always used in *every* circuit, they can be assumed.

2
BINARY CIRCUITS

This chapter features several projects to demonstrate some of the binary mathematical operations that often are needed within digital systems. If you aren't familiar with the binary system, don't worry about it. What you need to know will be explained with each project. In general, all you really need to know about the binary number system is that it has only two digits—1 and 0. In digital electronics a 1 is usually represented by a relatively HIGH voltage level (generally +VDD, or a little below), and a 0 is normally represented by a LOW voltage (generally ground potential). No other voltages are allowed.

DIGITAL COMPARATORS

One of the most useful analog circuits is the comparator, illustrated in Fig. 2-1. The name pretty much suggests the function. Two input signals are compared to one another. The output can take on one of three different values, depending on the relationship between the input values:

    +   A > B
    0   A = B
    −   A < B

Notice that there are no other possible combinations.

You're probably wondering why I'm talking about an analog circuit in a book of digital projects. The comparator is such a useful function, that it makes sense for digital

23

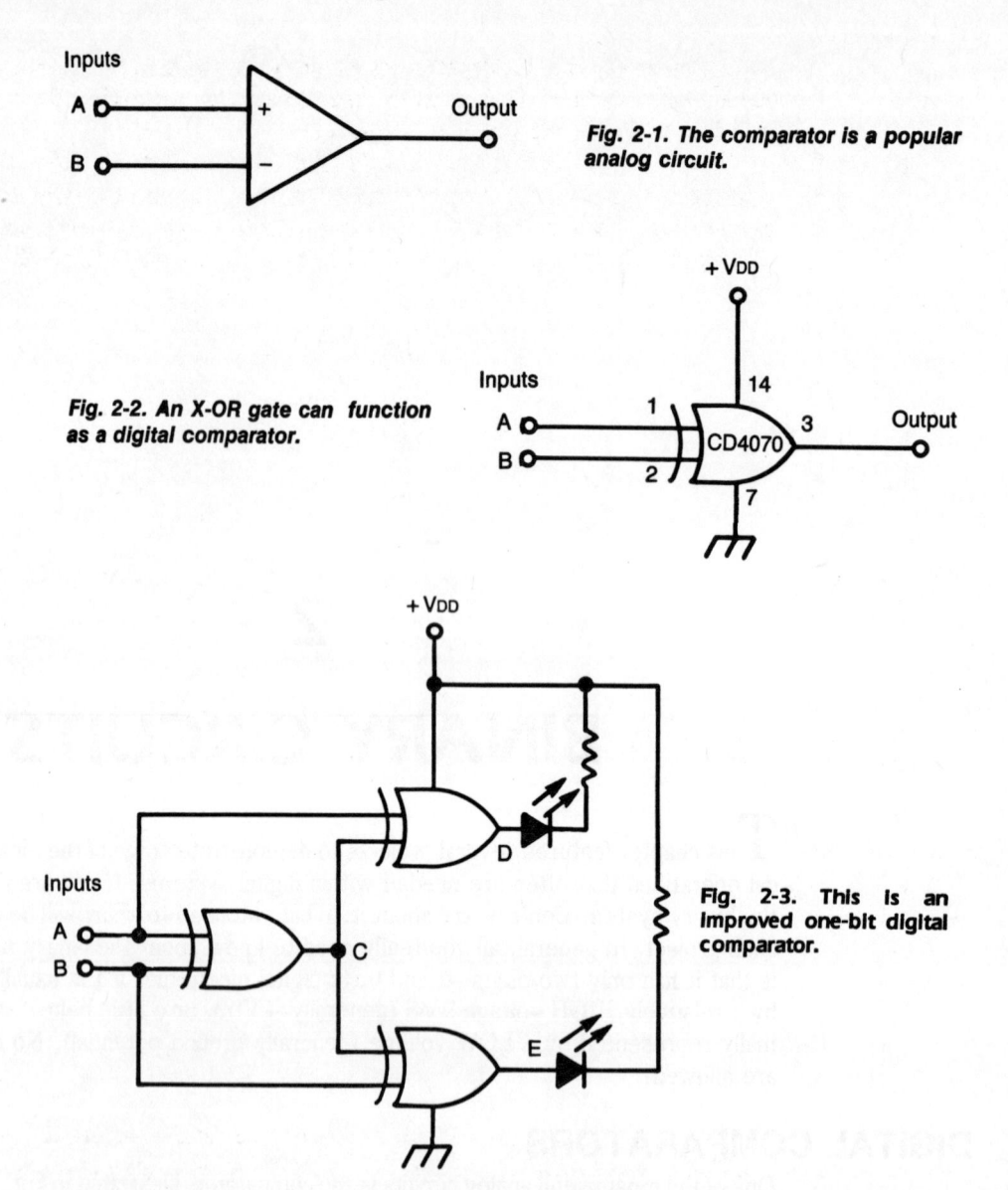

Fig. 2-1. The comparator is a popular analog circuit.

Fig. 2-2. An X-OR gate can function as a digital comparator.

Fig. 2-3. This is an improved one-bit digital comparator.

values too. Often it would be useful to have a circuit that could determine which of two digital inputs is the larger, that is, a digital comparator.

A super-simple one-bit digital comparator is shown in Fig. 2-2. This is nothing more than an exclusive-OR (X-OR) gate. As you should recall, an *X-OR* gate is sometimes called a difference detector. The output goes HIGH if the two inputs are at opposite states. That is, if one is LOW and the other is HIGH. If the two inputs are equal, (both LOW, or both HIGH) the output will be LOW.

This simple comparator does not indicate which of the two inputs is larger, but it does indicate whether or not they are equal.

24

An improved one-bit digital comparator circuit is illustrated in Fig. 2-3. Here we use three X-OR gates. This circuit will indicate equality, or which of the two bits is larger (which is a logic 1 and which is a logic 0).

To see how this circuit works let's take a look at its truth table. An intermediate point "c" is within the circuit, the inputs are A and B, and the outputs are D and E. In operation, the output states are indicated by the two LEDs. The appropriate LED is lit when its associated output is HIGH. Here is the complete truth table in Table 2-1:

INPUTS A B	c	OUTPUTS D E
0 0	0	0 0
0 1	1	1 0
0 1	1	0 1
1 1	0	1 1

Table 2-1.

There are four possible output combinations that can be indicated by the LEDs. They unambiguously indicate the relative values of the inputs:

Both LEDs dark	A = B = 0
Only LED D lit	A < B (A = 0, B = 1)
Only LED E lit	A > B (A = 1, B = 0)
Both LEDs lit	A = B = 1

If you look back over the operation of this circuit, something should look wrong about this circuit. It is absolutely useless. The two outputs don't give us any more information than we could get by connecting LEDs to the two original inputs. A one-bit digital comparator is too simple to be worthwhile. So why did we waste our time with this exercise in futility? For two reasons. First, this is a concrete example of *"Don't overlook the obvious."* Sometimes we can get ahead of ourselves working out a complex gating network when none is needed.

The other reason that I included this circuit is it helps explain the operation of multi-bit digital comparators, which are useful circuits.

A single bit is rarely useful by itself. It is exceedingly limited in the amount of information it can carry. If you are dealing with anything more complex than simple yes/no conditions, one bit just won't be enough.

Digital bits can be combined to form strings, which when taken together can take on complex coded meanings. The number of possible combinations increases exponentially as the number of bits in the string increases. All of the bits in the string are usually transmitted through the system together (in parallel).

A few common bit-string lengths are often used. Not surprisingly, since digital systems use the binary numbering system, the common lengths are powers of two. Specialized names are given to these strings demonstrated in Table 2-2:

NUMBER OF BITS	NAME	POSSIBLE COMBINATIONS
4	nibble	16
8	byte	256
16	word	65,536

Table 2-2.

As you can see, there are enough possible combinations to make a multi-bit digital comparator useful.

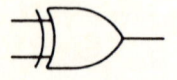

Project 1: Multi-bit Digital Comparator

A multi-bit digital comparator can be designed around individual gates, but the circuit would be terribly complicated. Some specialized ICs are available for comparator applications. One such device is the 74C85. By itself, this chip can compare two four bit nibbles. But it is even more powerful and versatile than that. Two (or more) 74C85s can be used together to compare longer digital strings. A two stage circuit for comparing a pair of eight-bit bytes, and the parts list for this project is given in Fig. 2-4.

Some explanation is needed for feeding in the inputs. Each eight-bit byte is broken up into two four-bit nibbles. X(L) is the least significant nibble, and X(H) is the most significant bit. Each byte is arranged like this:

$$A = A(H)\ A(L)$$
$$B = B(H)\ B(L)$$

For example:

$$A = 10100011 \quad A(L) = 0011 \quad A(H) = 1010$$
$$B = 11011001 \quad B(L) = 1001 \quad B(H) = 1101$$

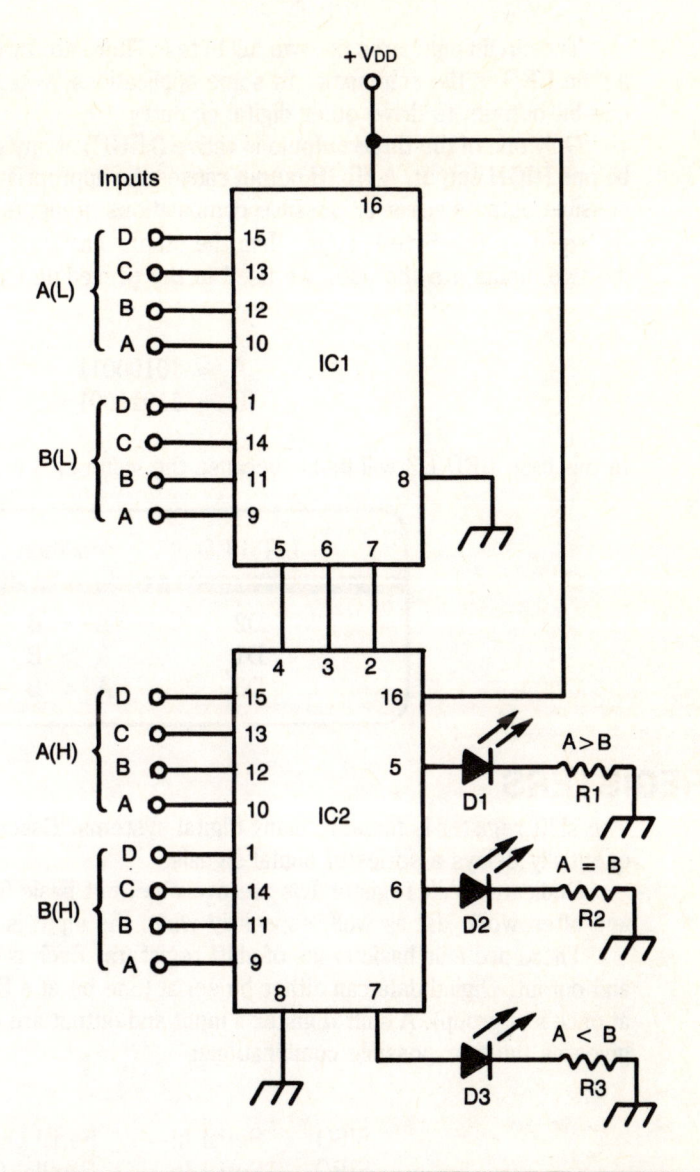

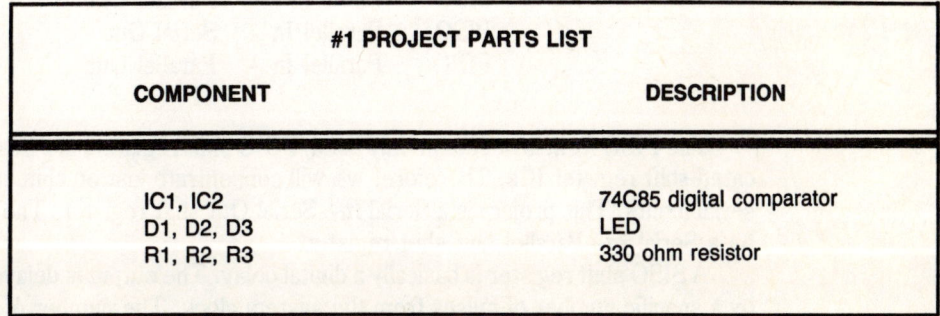

Fig. 2-4. This circuit compares two eight-bit bytes.

#1 PROJECT PARTS LIST	
COMPONENT	DESCRIPTION
IC1, IC2	74C85 digital comparator
D1, D2, D3	LED
R1, R2, R3	330 ohm resistor

The circuit compares the two full bytes. There are three outputs, each shown driving an LED in the schematic. In some applications, you might want to use the three one-bit outputs to drive other digital circuitry.

Only one of the three outputs is active (HIGH) at any given time. Always there will be one HIGH output. A HIGH output causes the appropriate LED to light up. The three possible outputs cover all possible combinations of inputs. Either the two input bytes are equal, or one byte is larger than the other. This covers all possibilities. Assuming the two inputs are the ones we used in the preceding example:

$$A = 10100011$$
$$B = 11011001$$

In this case LED D3 will be lit, because the value of A is smaller than the value of B.

LIT LED	condition
D2	A = B
D1	A > B
D3	A < B

SHIFT REGISTERS

The shift register is found in many digital systems. Essentially, it is a circuit that sequentially delays a series of digital signals.

Dedicated shift register ICs are available, but basic flip-flops tend to be cheaper, and often work just as well, especially when the input is to be fed in serially.

There are four basic types of shift registers. Each is defined by the type of input and output. Digital data can either be serial (one bit at a time) or parallel (several bits at once in a group). A shift register's input and output are either serial or parallel. This gives us the four possible combinations:

SISO	Serial In—	Serial Out
SIPO	Serial In—	Parallel Out
PISO	Parallel In—	Serial Out
PIPO	Parallel In—	Parallel Out

The PISO form is not commonly used. PIPO shift registers are best built from dedicated shift register ICs. Therefore, we will concentrate just on shift registers that use serial inputs. This project is a Serial In—Serial Out shift register. The next project will be a Serial In—Parallel Out shift register.

A SISO shift register is basically a digital delay. The output is delayed from the input by a specific number of pulses from the system clock. The number of delay pulses will be equal to the number of stages in the shift register circuit.

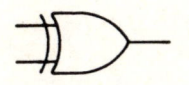

Project 2: SISO Shift Register

A block diagram for a simple SISO shift register is shown in Fig. 2-5. Notice that the shift register is nothing more than a string of D-type flip-flops, each one triggering the next one in line. Four stages are shown.

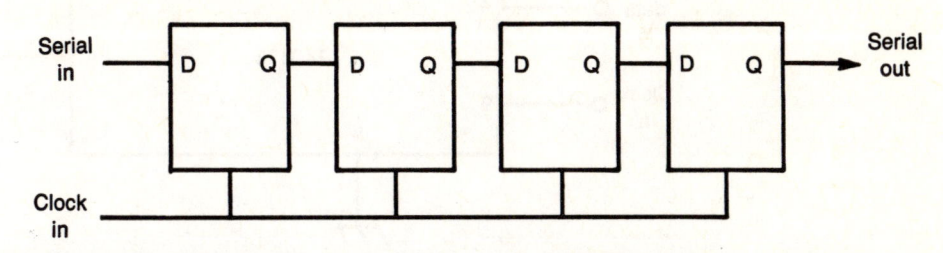

Fig. 2-5. A SISO shift register is made up of a string of flip-flops.

To understand the workings of the SISO shift register, let's follow a typical example. We will assume that initially the shift register is filled with 0s (logic LOW held by all four stages). Then we will input the following string of data:

$$1-1-0-1-1-1-0-0-1-0$$

This pattern will be followed by a string of 0s to clear out the shift register.

The shift register accepts one new input bit on each clock pulse. When we begin, the shift register holds 0000.

Here are the inputs and outputs for each clock pulse in Table 2-3:

PULSE #	INPUT	OUTPUT	
1	1	0	
2	1	0	
3	0	0	
4	1	0	
5	1	1	(data out begins)
6	1	1	
7	0	0	
8	0	1	
9	1	1	
10	0	1	
11	0	0	
12	0	0	
13	0	1	
14	0	0	
15	0	0	(register cleared)

Table 2-3.

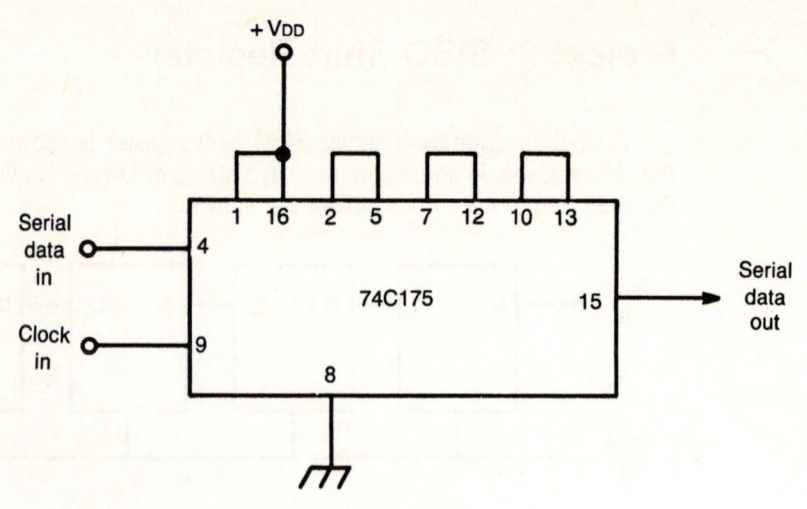

Fig. 2-6. This is a practical SISO shift register circuit.

The output is always four pulses behind the input.

A practical SISO shift register circuit is made up of the four sections of a 74C175 quad D-type flip-flop shown in Fig. 2-6. Four independent D-type flip-flops are contained in a single 74C175 chip. Since no external components are required for this circuit, no parts list is given for this project.

You can observe the output state with the LED display illustrated in Fig. 2-7. The resistor's purpose is to drop the current flowing through the LED to a safe value. The resistance should be between 250 Ω and 1000 Ω. The smaller the dropping resistor, the brighter the LED will glow.

A manual input switch network for manually entering data is shown in Fig. 2-8.

While the circuit presented here is just a demonstration of the SISO shift register, it can be put to practical work in projects of your own.

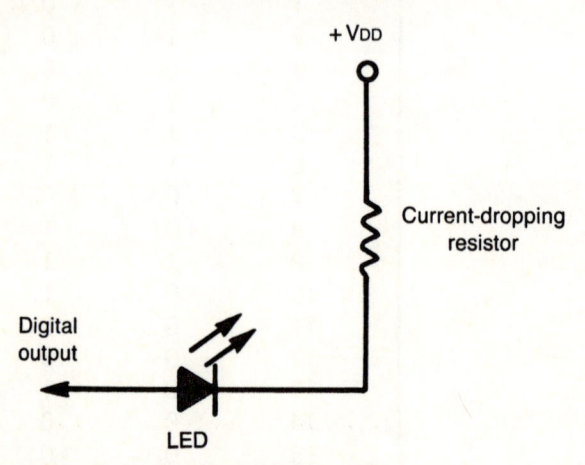

Fig. 2-7. The output state can be indicated with an LED.

Switch open = 0
Switch closed = 1

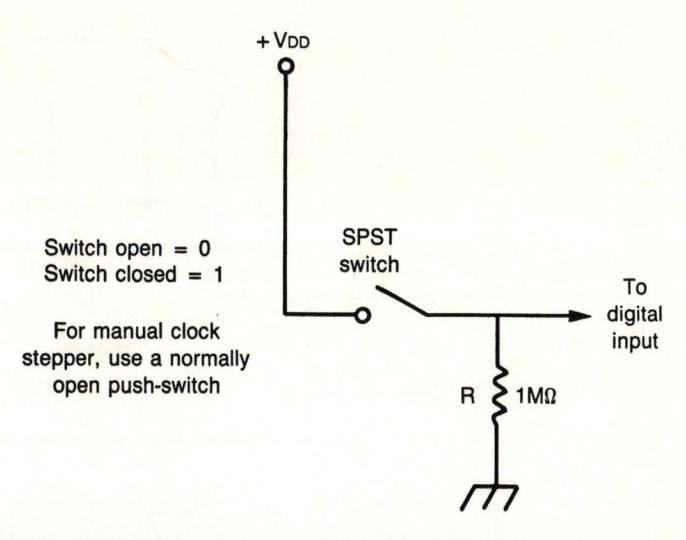

For manual clock stepper, use a normally open push-switch

Fig. 2-8. An SPST switch can be used to manually enter data.

Project 3: SIPO Shift Register

A SIPO (Serial In—Parallel Out) shift register is very similar to the SISO (Serial In—Serial Out) type. The only real difference is that the SISO shift register has only a single output, after the final flip-flop stage, while the SIPO shift register has an output after each individual flip-flop stage.

A block diagram of a four stage SIPO shift register is shown in Fig. 2-9.

Notice that the SIPO shift register is identical to the SISO shift register, except additional outputs are tapped off from internal circuit connection points.

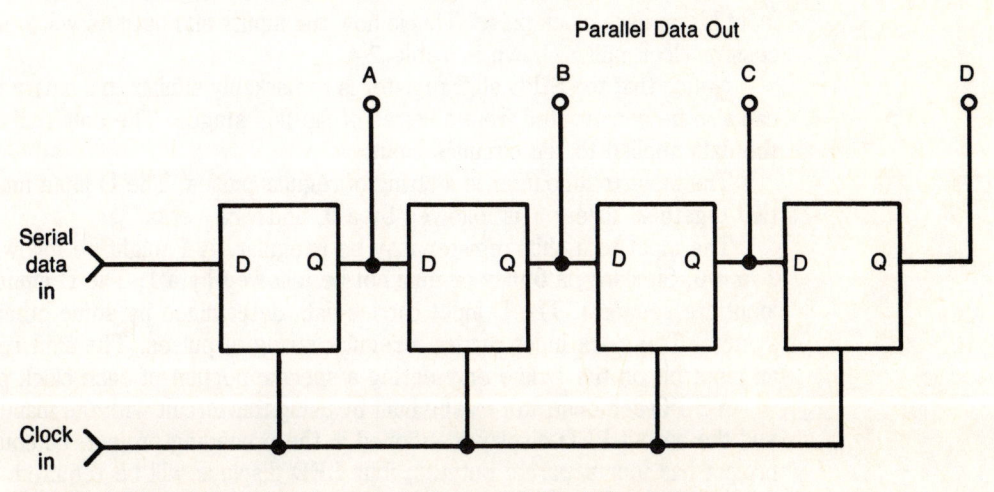

Fig. 2-9. A SIPO shift register is quite similar to a SISO shift register.

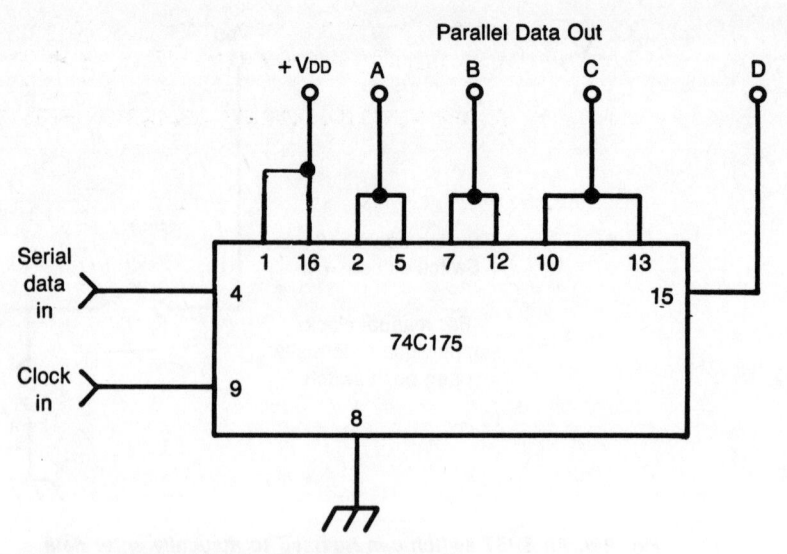

Fig. 2-10. This is a practical SIPO shift register circuit.

A practical SIPO circuit is shown in Fig. 2-10. Once again we are using the 74C175 quad D-type flip-flop IC. Except for the additional outputs, this circuit is the same as the SISO shift register circuit presented in the last project.

The input data is still entered serially. Each bit is passed from output to output in sequential fashion. Let's consider a typical example. Initially the shift register is assumed to be cleared, that is, each flip-flop stage holds a 0. All outputs are LOW. Then the following input pattern is entered serially:

$$1 - 1 - 0 - 1 - 0 - 0 - 1 - 1 - 1 - 0 - 1$$

The input pattern is followed by a string of 0s to clear the shift register.

As with the SISO shift register, the SIPO shift register's serial input accepts one input bit on each clock pulse. This is how the inputs and outputs will look for each successive clock pulse shown in Table 2-4.

Notice that the SIPO shift register is remarkably similar to a binary counter, which can also be constructed from a series of flip-flop stages. The only real difference is in the data applied to the circuit's input.

The input to a counter is a string of regular pulses. The D input and the clock are tied together. Every 1 is followed by a 0, and vice versa.

The input to a shift register may be irregular. A 1 might be followed by either a 1 or a 0. Similarly, a 0 may or may not be followed by a 1. The D input and the clock input are separate. The D input carries data determined by some other circuit in the system. The clock input carries a regular string of pulses. The shift register accepts an input bit on the D line only during a specific portion of each clock pulse.

A SIPO demonstrator can be built by using this circuit with the manual input switch and the output LED display mentioned in the preceding project. Of course, since this project has four separate outputs, four LED displays will be required.

CLOCK PULSE #	INPUT	OUTPUTS A B C D	
1	1	0 0 0 0	
2	1	1 0 0 0	
3	0	1 1 0 0	
4	1	0 1 1 0	
5	0	1 0 1 1	
6	0	0 1 0 1	
7	1	0 0 1 0	
8	1	1 0 0 1	
9	1	1 1 0 0	
10	0	1 1 1 0	
11	1	0 1 1 1	
12	0	1 0 1 1	
13	0	0 1 0 1	
14	0	0 0 1 0	
15	0	0 0 0 1	
16	0	0 0 0 0	(cleared)
17	0	0 0 0 0	

Table 2-4.

BINARY ADDITION

The term *digital* implies mathematical operations. What is a computer but a superpowerful binary adding machine?

Digital electronics use binary numbers, rather than the digital numbers we are more familiar with. The decimal number system has ten digits:

$$0 - 1 - 2 - 3 - 4 - 5 - 6 - 7 - 8 - 9$$

If we need to express a value higher than 9, we use an additional digit to the immediate left of the original digit column. Digits in this new column have their value multiplied by the system base, or ten. For example:

$$96 = 6 + (9 \times 10)$$

This principle can be expanded indefinitely. Each time we add a new digit column to the left that digit's value is raised by the next power of ten. For example:

$73156 =$
$\qquad 6 + (5 \times 10) + (1 \times 10^2) + (3 \times 10^3) + (7 \times 10^4) =$
$\qquad 6 + (5 \times 10) + (1 \times 100) + (3 \times 1000) + (7 \times 10000)$

The binary system works in the same way, except the base is two. There are only two digits available in the binary number system:

0 and 1

If you want to express a value larger than one, then you have to use an additional digit column. Each new column in the binary system has its value raised by the next power of two. For example:

$$\begin{aligned} 10011 &= 1 + (1 \times 2) + (0 \times 2^2) + (0 \times 2^3) + (1 \times 2^4) \\ &= 1 + (1 \times 2) + (0 \times 4) + (0 \times 8) + (1 \times 16) \\ &= 1 + 2 + 0 + 0 + 16 \\ &= 19 \end{aligned}$$

10011 in binary equals 19 in decimal.

The following in Table 2-5 is a comparison of binary and decimal numbers:

BINARY		DECIMAL
0000	=	0
0001	=	1
0010	=	2
0011	=	3
0100	=	4
0101	=	5
0110	=	6
0111	=	7
1000	=	8
1001	=	9
1010	=	10
1011	=	11
1100	=	12
1101	=	13
1110	=	14
1111	=	16
10000	=	17

Table 2-5.

and so on.

Notice that leading zeroes are normally used in writing binary numbers. Binary digits are usually separated into groups of three or four, to make them easier to read. A binary digit is usually called a *bit*. **BI**nary digi**T**. Get it? Now you know where they got that silly word.

The binary system is very awkward for people to use, but it is perfect for digital circuits. Remember that a digital signal can have either of two unambiguous values—0 or 1.

Any digital computer or calculator actually performs all mathematical operations in the binary system.

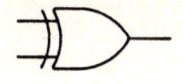

Project 4: Binary Adder

A very simple, limited, but functional binary calculator circuit is shown with the parts list for this project in Fig. 2-11.

This circuit uses all four NAND gates in a CD4011 IC, along with an additional inverter. A CD4049 hex inverter is listed in the parts list, but if you prefer, you can use one section of the second CD4011 quad NAND gate. Just short the inputs together.

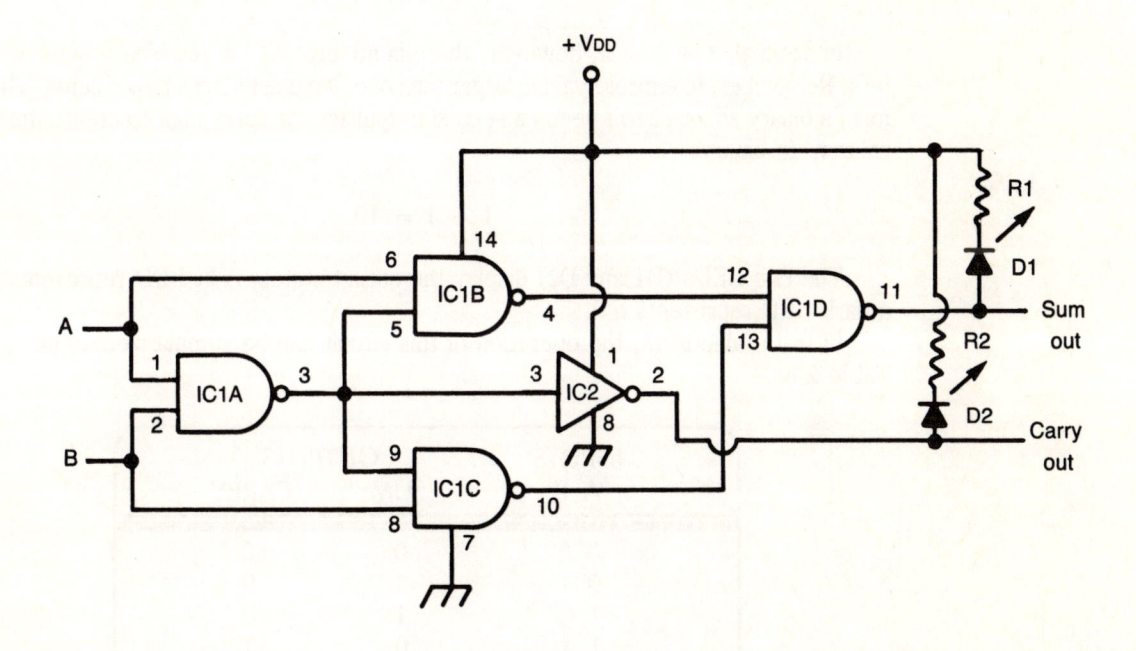

#4 PROJECT PARTS LIST

COMPONENT	DESCRIPTION
IC1	CD4011 quad NAND gate
IC2	CD4049 hex inverter (see text)
D1, D2	LED
R1, R2	330 ohm resistor

Fig. 2-11. This is the circuit for a simple binary adder.

This circuit is primarily intended for demonstration purposes, because it really doesn't do much but illustrate the basics of binary mathematics. With a little imagination, you could expand this circuit for more complex functions.

The circuit adds two single digit binary numbers, with two one-bit inputs there are four possible input combinations:

$$0 + 0; 0 + 1; 1 + 0; 1 + 1$$

The results for the first three combinations should be obvious, since they are the same as in the decimal system:

$$0 + 0 = 0$$
$$0 + 1 = 1$$
$$1 + 0 = 1$$

In decimal, $1 + 1 = 2$, however, there is no digit "2" in the binary number system. Remember, to express values larger than one, we must start a new column. Therefore, a binary adder circuit needs a second output for the carry digit to create the new column. In binary:

$$1 + 1 = 10$$

The two LEDs (D1 and D2) display the output states. A lit LED represents a 1, a dark LED represents a 0.

In truth table form, the operation of this circuit can be summarized as follows in Table 2-6:

INPUTS A B	OUTPUTS SUM	CARRY
0 0	0	0
0 1	1	0
1 0	1	0
1 1	0	1

Table 2-6.

This circuit is called a binary half-adder because it does not have a carry input from a previous column. As an exercise, try designing a two digit binary full-adder. Remember, the circuitry for the second column must handle *three* inputs:

A's most significant bit
B's most significant bit
The carry from least significant bits

Believe it or not, this simple circuit (or something very similar) is at the heart of the most sophisticated digital computer in the world.

MAJORITY LOGIC CIRCUITS

In most digital gating applications, the output (or outputs) state is determined by very specific patterns of input states. In some cases, however, exact patterns of individual inputs being at specific states is not as important as broader, more generalized patterns.

As an example, consider a voting machine network. Multiple inputs are all given equal weights. The actual state of input C, for example, isn't important. We are only interested in which input state is dominant over the entire set of inputs. In other words, rather than direct, fixed gating logic, we need a system of "majority rule." Not surprisingly, circuits that work along these lines are known as *majority logic* circuits.

Dedicated majority logic chips are available; however, they can often be fairly difficult to locate. Besides, it is more educational to use standard gates, so the hobbyist can better determine what is happening within the circuit. This is the approach we will use in this demonstration project. The majority logic gating circuit can be used in many practical applications.

A majority logic gating circuit allows the inputs to "vote" on the desired output condition. In a non-inverted circuit, if the majority of inputs are HIGH, the output will be HIGH. If the majority of inputs are LOW, the output will be LOW. To avoid problematic tie votes, there should be an odd number of inputs.

The smallest practical number of inputs to a majority logic gating circuit is three. For the output to be HIGH, at least two of the inputs must be HIGH. Notice that for a three input gate, there are eight possible input bit combinations (ranging from 000 to 111). In majority logic this is effectively reduced to just four possibilities:

All three inputs LOW
Any two inputs LOW
Any two inputs HIGH
All three inputs HIGH

Notice that all possible bit combinations are covered here.

The possible combinations can be reduced to two. If all three inputs are the same, at least two must be identical. In other words, either:

Two or more inputs are LOW

or:

Two or more inputs are HIGH

One (and only one) of these two statements must be true for all possible input bit combinations.

The truth table for a three-input Majority Logic gating circuit is shown in Table 2-7:

37

INPUTS A B C	OUTPUT
0 0 0	0
0 0 1	0
0 1 0	0
0 1 1	1
1 0 0	0
1 0 1	1
1 1 0	1
1 1 1	1

Table 2-7.

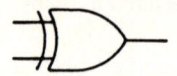

Project 5: Majority Logic Demonstrator

Any input/output pattern can be generated by combining standard logic gates, so there is no reason we can't create a majority logic circuit from scratch. As always, there are many possible ways to generate the desired input/output pattern, but the most direct method is the one illustrated in Fig. 2-12.

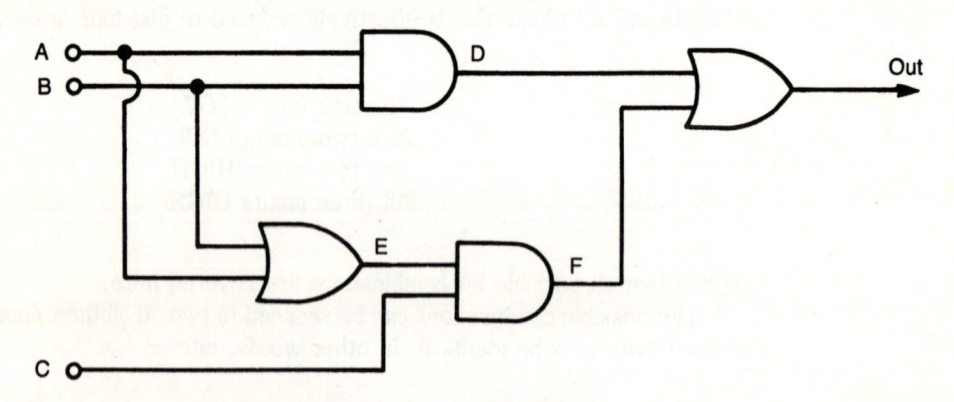

Fig. 2-12. This is a simple three-input majority logic circuit.

Operation of this circuit is made clear by the complete truth table shown in Table 2-8, which lists the input and output states, as well as the states at each of the intermediate points within the circuit.

As the number of inputs is increased, the majority logic gating circuit becomes more complicated. Figure 2-13 shows a practical five-input majority logic network. A total of six AND gates, and five OR gates are used in this circuit. IC1 and IC3 are quad AND gate ICs, such as the CD4081B. IC2 and IC4 are quad OR gate packages, such as the CD4071B. Notice that there are unused gates in IC3 and IC4. The truth table for this circuit is in Table 2-9.

| INPUTS | OUTPUT |
A B C d e f	
0 0 0 0 0 0	0
0 0 1 0 1 0	0
0 1 0 0 1 0	0
0 1 1 0 1 1	1
1 0 0 0 1 0	0
1 0 1 0 1 1	1
1 1 0 1 1 0	1
1 1 1 1 1 1	1

Table 2-8.

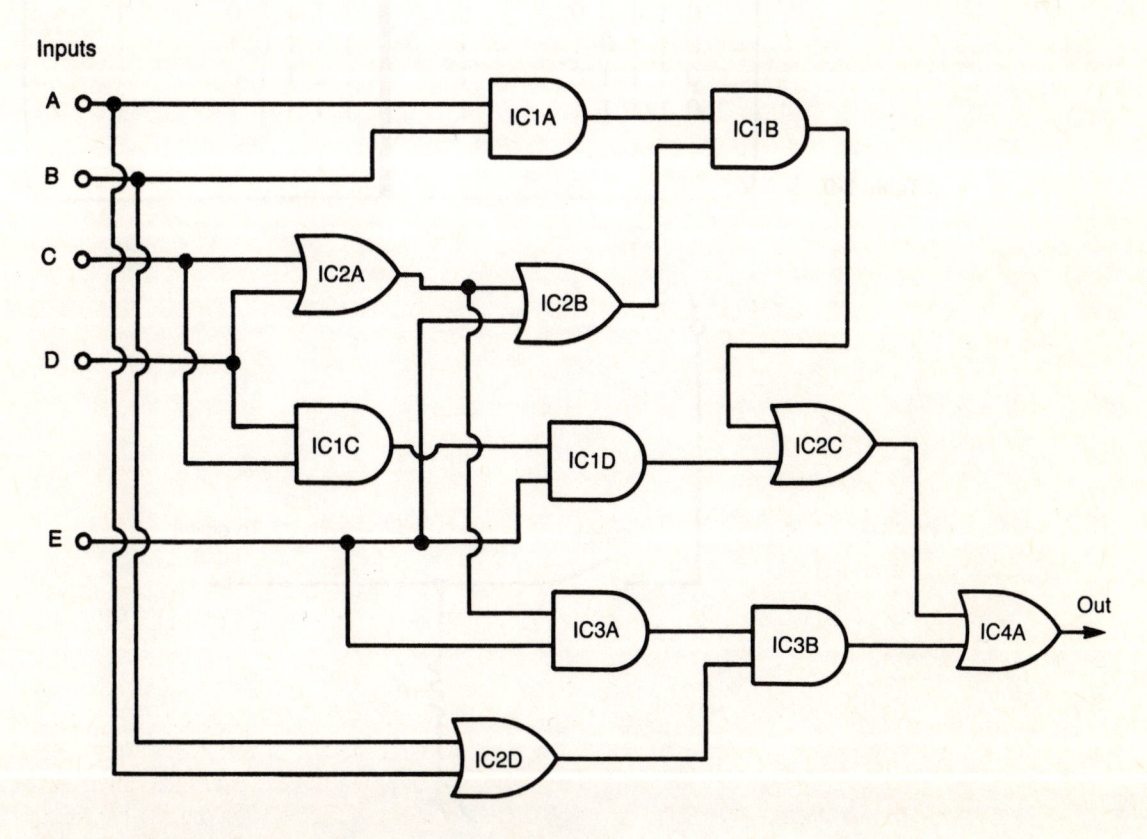

Fig. 2-13. This is a five-input majority logic demonstrator.

INPUTS A B C D E	OUTPUT	INPUTS A B C D E	OUTPUT
0 0 0 0 0	0	1 0 0 0 0	0
0 0 0 0 1	0	1 0 0 0 1	0
0 0 0 1 0	0	1 0 0 1 0	0
0 0 0 1 1	0	1 0 0 1 1	1
0 0 1 0 0	0	1 0 1 0 0	0
0 0 1 0 1	0	1 0 1 0 1	1
0 0 1 1 0	0	1 0 1 1 0	1
0 0 1 1 1	1	1 0 1 1 1	1
0 1 0 0 0	0	1 1 0 0 0	0
0 1 0 0 1	0	1 1 0 0 1	1
0 1 0 1 0	0	1 1 0 1 0	1
0 1 0 1 1	1	1 1 0 1 1	1
0 1 1 0 0	0	1 1 1 0 0	1
0 1 1 0 1	1	1 1 1 0 1	1
0 1 1 1 0	1	1 1 1 1 0	1
0 1 1 1 1	1	1 1 1 1 1	1

Table 2-9.

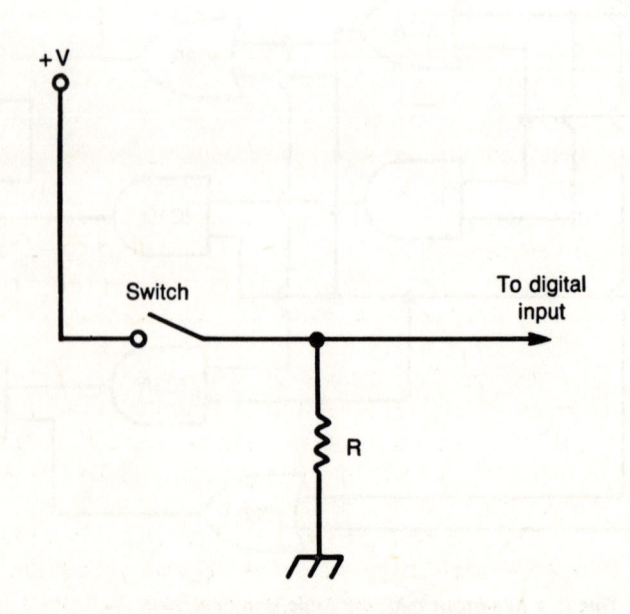

Fig. 2-14. An SPST switch can be used to manually enter data.

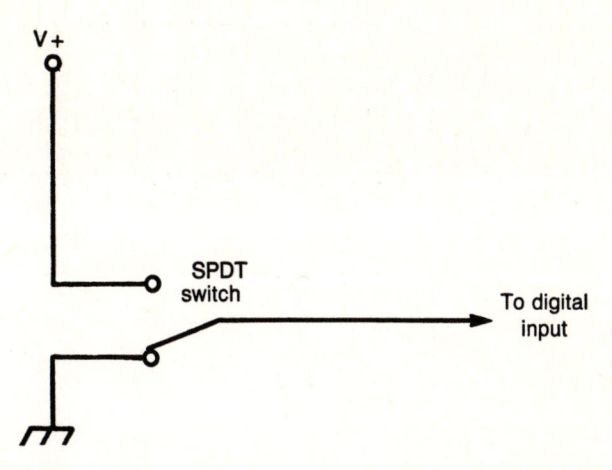

Fig. 2-15. Alternatively, an SPDT switch can be used to manually enter data.

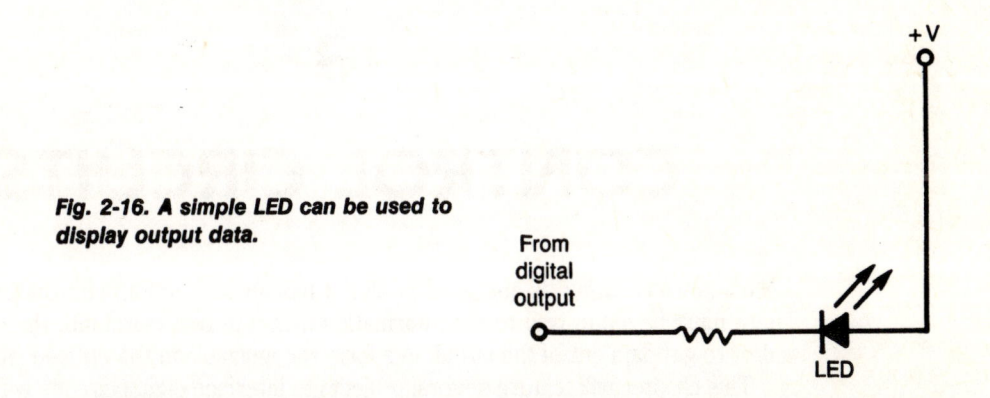

Fig. 2-16. A simple LED can be used to display output data.

To demonstrate the majority logic operation, just add a simple switch to each input. The switching circuit is shown in Fig. 2-14. A simple SPST switch is used. When the switch is open, the appropriate input is grounded through resistor R (logic 0). Closing the switch pulls the appropriate input up to V+ (logic HIGH). Resistor R isolates the input from ground when the switch is closed. This resistor should have a fairly high value. Something in the 1 Megohm to 10 Megohm range is recommended, although the exact value isn't important. You just need a large resistance here to prevent a short circuit effect.

If you prefer, you can use SPDT switches for the inputs, wired as illustrated in Fig. 2-15.

A simple LED can be used to indicate the output state. It should be wired into the circuit as shown in Fig. 2-16. The resistor is included to limit the current through the LED to a safe value. Its value should be in the 250 ohm to 1000 ohm (1kΩ) range. The smaller this resistance, the brighter the LED will glow. I usually use a 330 ohm or 470 ohm resistor in such applications.

The input switches and the output indicator are shown separately in this project to make it more convenient for you to utilize the majority logic circuit in other projects of your own. Any digital inputs can be used, and of course, the output signal can be used in the same way as any other digital signal.

3
CONTROL CIRCUITS

No electronic circuit is any good at all if it has no connection with the outside world. There must be some way to get information from the real world into the circuit (input) and/or to get data out of the circuit in a form recognizable in the outside world (output).

This chapter will feature several projects to interface digital circuits without the outside world. Some of the projects accept some kind of non-digital physical input, while others have a direct effect on some non-digital output device.

TOUCH SWITCHES

Touch switches are usually popular projects. Any electrical or electronic device can be activated by a light touch of a fingertip.

The touch switch itself is nothing more than a pair of simple conductive plates. Small pieces of copper-clad circuit board will do nicely. There is a small space between the two touch plates. The plates are positioned so that they can easily be bridged by a fingertip, as shown in Fig. 3-1.

The fingertip provides a short circuit between the plates, effectively *closing the switch*.

IMPORTANT!!! Since the human body serves as part of the signal path, touch switches should be operated by battery power *only*. **Never use an ac power supply with any touch switch project.**

You might think that it would be perfectly safe to use a well regulated ac-to-dc power supply. After all, there will only be low-power dc voltages flowing in the touch circuit.

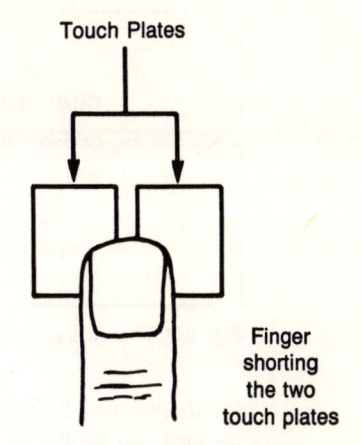

Fig. 3-1. In a touch switch, the user's finger completes the circuit.

Normally, yes. But there is always a chance for an unexpected short circuit that allows a high-voltage ac signal to reach the touch plates. *It is not worth the risk!* Use battery power only!

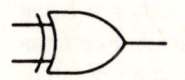

Project 6: Touch Switch

A simple touch switch circuit is shown in Fig. 3-2 along with the parts list.

Switch S1 is an enable switch. When this switch is open, pin #1 is grounded through resistor R3. This gives a logic LOW input. Since a NAND gate is used, the output will always be HIGH, regardless of the input state at pin #2. Closing switch S1 enables the

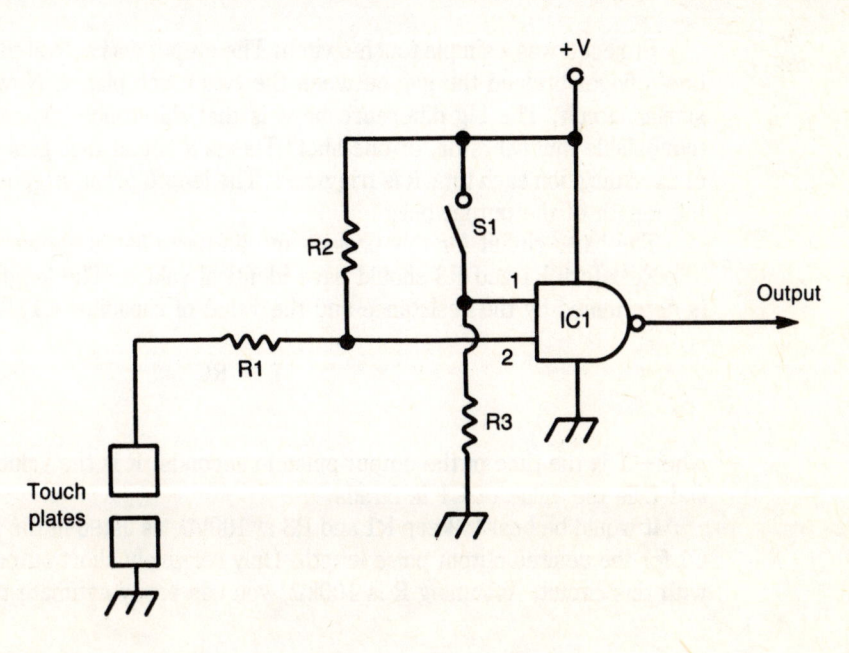

43

#6 PROJECT PARTS LIST	
COMPONENT	DESCRIPTION
IC1	CD4011 quad NAND gate
R1	100K resistor
R2, R3	10 Megohm resistor

Fig. 3-2. This is a practical touch switch circuit.

touch switch. The logic state at pin #2 will control the output at pin #3. The signal will be inverted by the gate.

Let's assume that S1 is closed, enabling the circuit. When the touch pads are not being touched pin #2 is held HIGH through resistor R2, the output will be LOW. Touching the touch pads creates a short through ground. Pin #2 is grounded through R1 which has a lower resistance than R2, so the input at pin #2 goes LOW, driving the output HIGH.

Anything that can be controlled by a logic signal can be controlled by this touch switch project. Note that the output will be HIGH only as long as the touch pads are being bridged with a finger. When the finger is removed the output goes LOW again. For some applications you might want to follow the basic touch circuit with some kind of latch, for an on/off touch type control.

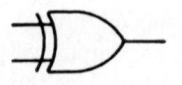

Project 7: Timed Touch Switch

Project 6 was a simple touch-switch. The output was activated only as long as someone's finger bridged the gap between the two touch plates. Now we will work with a similar circuit. The big difference here is that this touch switch circuit has a built-in monostable multivibrator, or one-shot. This is a circuit that generates an output pulse of fixed duration each time it is triggered. The length of the trigger pulse does not affect the length of the output pulse.

The basic circuit for this project and its parts list is shown in Fig. 3-3.

Resistors R1 and R3 should have identical values. The length of the output pulse is determined by the resistance and the value of capacitor C1. The rough formula is:

$$T = RC$$

where T is the time of the output pulse in seconds, R is the value of R3 (R1) in ohms, and C is the value of C1 in farads.

It would be best to keep R1 and R3 at 100kΩ, as listed in the parts list, and choose C1 for the desired output pulse length. Only relatively short output pulses are possible with this circuit. Assuming R is 100kΩ, you can easily estimate the value of C just by

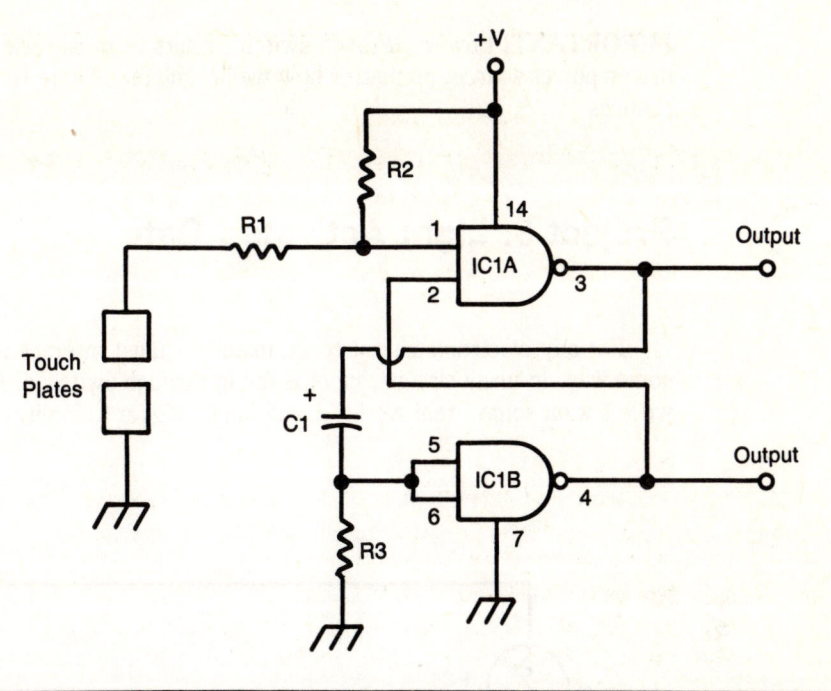

(schematic of touch switch circuit with IC1A and IC1B NAND gates, touch plates, R1, R2, R3, C1)

#7 PROJECT PARTS LIST

COMPONENT	DESCRIPTION
IC1	CD4011 quad NAND gate
R1, R3	100K resistor
R2	10 Megohm resistor
C1	2.2 µF electrolytic capacitor (see text)

Fig. 3-3. This touch switch has a built-in delay.

multiplying the desired time (in seconds) by 10. This will give you the approximate capacitance needed in microfarads (µF):

$$C = 10T$$

Using the component values given in the parts list will result in an output pulse a little over 2 seconds in length. For an output pulse of 1 minute (60 seconds), try combining a 470 µF capacitor and a 120 µF capacitor in parallel.

Remember that capacitors—especially large value electrolytic capacitors—have very wide tolerances, so the timing equations will not be exact. They should be close enough for most purposes.

IMPORTANT! Power *all* touch switch circuits from batteries only. Never use an ac-driven power source, no matter how well regulated. There is no sense in taking foolish chances.

Project 8: Light Activated Gate

For any electronic circuit to be useful, it must interact with the *outside world* in some way. In many circuits, input is fed in through switches. But in some applications, we will want some "real-world" condition to trigger a circuit. Figure 3-4 shows a circuit

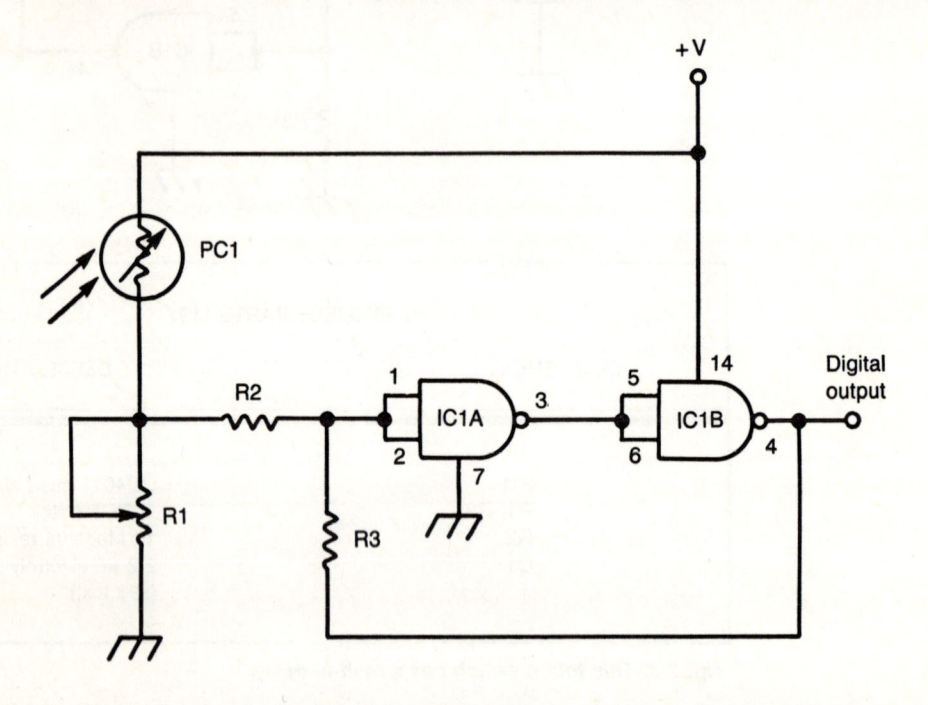

#8 PROJECT PARTS LIST

COMPONENT	DESCRIPTION
IC1	CD4011 quad NAND gate
PC1	photocell
R1	100kΩ potentiometer
R2	10kΩ resistor
R3	120kΩ resistor

Fig. 3-4. Light intensity determines whether the output of this circuit will be HIGH or LOW.

for triggering a CMOS circuit with a change in light level along with its parts list. The sensor is a simple resistive photocell. Potentiometer R1 is used to adjust the trip-point of the circuit, that is, how much light is required to trigger the circuit.

There are two ways to use this circuit. You could install the sensor in a dark area and shine a flashlight on it, or alternatively, the sensor could be placed in normal ambient light. It can then be triggered by placing your hand (or any other object that will cast a shadow) over the photocell. When the light striking the cell's surface drops below the level determined by the setting of R1, the circuit will be triggered.

There are many possible applications for this circuit. Some of the more obvious applications include detecting passing objects, or people passing through a doorway. This circuit is a real natural for use in burglar alarm systems.

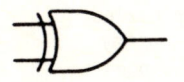

Project 9: Pulse Delayer

In some applications we may need to delay a pulse for a specific amount of time. That is, the pulse is used by one part of the circuit, then later the same pulse is used by another part of the circuit.

There are several possible approaches to this problem. One of the simplest is to use a clocked counter. A low frequency clock is used. The pulse is delayed for however long it takes the counter to reach the specified count.

A circuit of this type is shown in Fig. 3-5, along with the parts list.

IC1 is just a 7555 timer wired in the astable mode. (You can use a standard 555 timer, if you prefer.) Potentiometer R1 is used to adjust the clock frequency. In some applications, such adjustability could be unnecessary, and possibly even undesirable. In this case, R1 and R2 can be combined into a single fixed resistor of the appropriate value for the desired clock frequency.

A fairly large timing capacitor (C1) is used to slow down the clock rate. Increasing the value of this capacitor will decrease the clock frequency.

Switch S1 is a Normally Open SPST push-switch. It is used to manually trigger the delay circuit. In most practical applications you will want to eliminate S1 and R4, and use a digital signal from another part of the circuit as the trigger.

When S1 is open, pin #15 of IC2 is held LOW because it is grounded through resistor R4. Closing the switch directly connects the positive supply voltage to pin #15 of IC2 momentarily. This acts as a HIGH pulse, triggering the circuit.

IC2 is a CD4017 decimal counter. It has ten outputs—one for each decimal digit (0 through 9). At any given instant, one and only one of the ten outputs will be HIGH. The other nine will be LOW. The chip's ENABLE input (pin #13) is shorted to the highest output count 9 (on pin #11). When you first power up the circuit, the counter will start at 0 and count up to 9, one step per clock pulse. When a count of 9 has been reached, the counter will DISABLE itself and stop, holding the 9 output HIGH. Nothing more will happen until IC2 receives a trigger pulse at pin #15. This pin is the counter's RESET. When activated, the counter will be reset back to 0. The chip will now be ENABLEd.

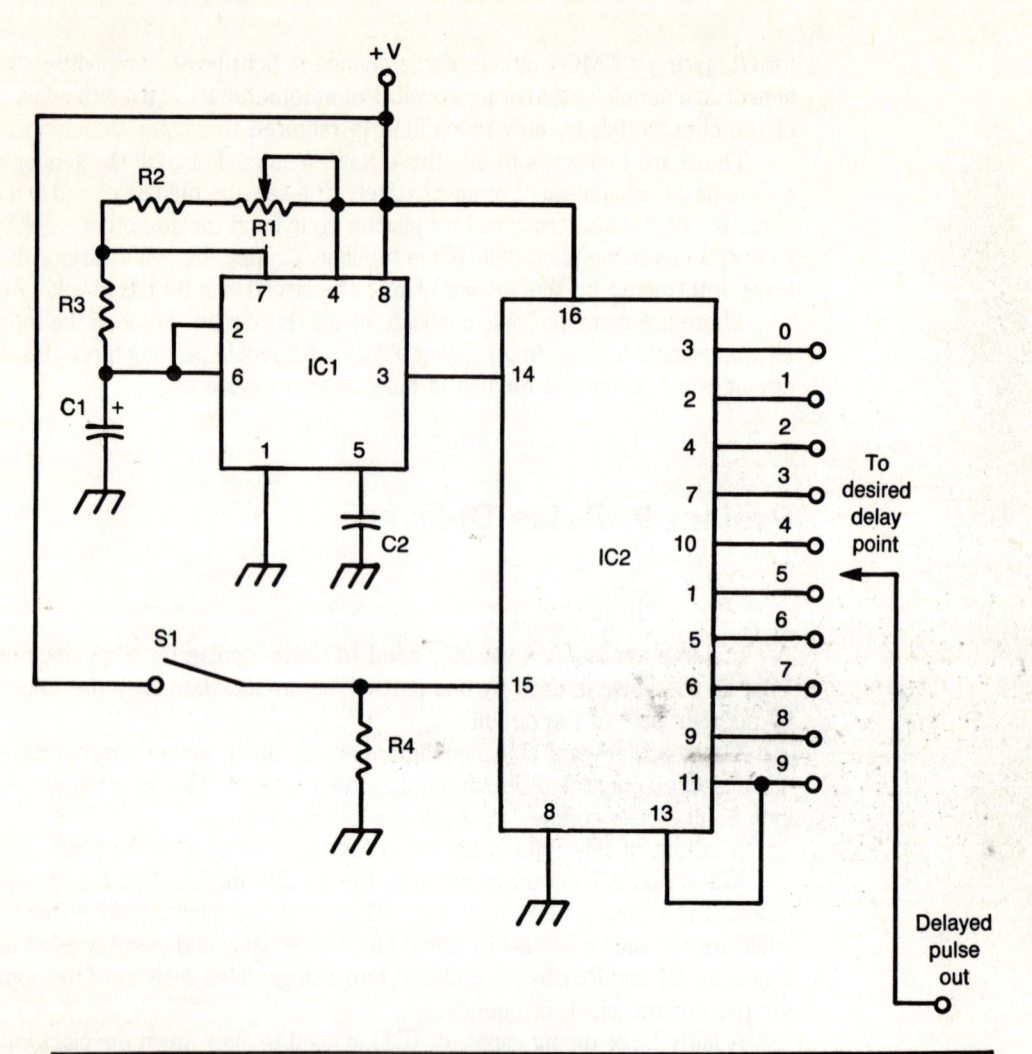

Fig. 3-5. This circuit produces an output pulse in response to a trigger pulse, after a user-determined delay.

#9 PROJECT PARTS LIST	
COMPONENT	DESCRIPTION
IC1	7555 timer
IC2	CD4017 decimal counter
C1	100 µF electrolytic capacitor (see text)
C2	0.01 µF capacitor
R1	100kΩ potentiometer
R2, R3	47kΩ resistor
R4	1MΩ resistor

The counter will bring each of its outputs HIGH one by one in sequence, at a rate determined by the clock frequency. To get the delayed pulse, just tap off from the appropriate counter output. The delay will be equal to:

$$DELAY = COUNT \times T$$

where T is the length of a single complete clock pulse.

For example, let's say the clock has a frequency of 0.1 Hz. The pulse time is the reciprocal of the frequency:

$$\begin{aligned}T &= 1/F \\ &= 1/0.1 \\ &= 10 \text{ seconds}\end{aligned}$$

The delay at each output is shown in Table 3-1:

OUTPUT	PIN #	DELAY
0	3	0T = 0 seconds
1	2	1T = 10 seconds
2	4	2T = 20 seconds
3	7	3T = 30 seconds
4	10	4T = 40 seconds
5	1	5T = 50 seconds
6	5	6T = 60 seconds
7	6	7T = 70 seconds
8	9	8T = 80 seconds
9	11	9T = 90 seconds

Table 3-1.

That's a pretty good-sized range for a single circuit. Multiple outputs can be used for different delays. For example, you might use pin #4 for a delay of 20 seconds to control one digital circuit, and drive a second circuit from pin #1 for a delay of 50 seconds.

Even though there are ten actual outputs from the CD4017, for most applications, you will really have a choice of only eight practical outputs. Output 0 (pin #3) is useless for this purpose, since it introduces no delay at all. At the other extreme, output 9 (pin #11) is potentially problematic, because this output is held HIGH the entire time the counter is inactive. It only goes LOW from the time that the counter is triggered until a count of 9 is reached. In other words, this output is LOW for a period equal to 8T from the instant the circuit is triggered. This may or may not cause problems in your particular applications. In a few applications, it might even be helpful. The important thing to remember is that output 9 operates differently from the other delay outputs.

For demonstration purposes, you can use an LED on one or more of the outputs. Be sure to use a current dropping resistor, as shown in Fig. 3-6.

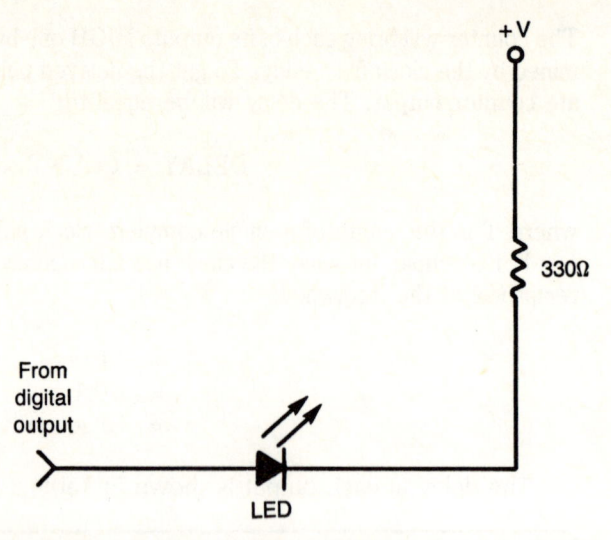

Fig. 3-6. A current dropping resistor protects an LED being used to display an output state.

For comparison purposes, you might want to monitor the clock's output with another LED connected to pin #3 of IC1.

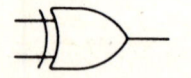

Project 10: Automatic Night Light

Have you ever come home late at night when you've forgotten to turn on the porch light? Isn't it frustrating trying to fit the key into the lock in the dark? Or have you ever gotten up in the middle of the night to go to the bathroom, and tripped over the cat, or one of the kids' toys. Then you have to fumble around for the light switch.

A small night light can make life easier, but it's ridiculous (and wasteful) to let it burn during daylight hours. But who can remember to turn the darn thing on every evening and off every morning? This is an ideal job for automation.

You can rig up a timer of some sort, but in this case that would be overkill. The exact time isn't important. When it's dark we want the light to come on. When there is sufficient light, the night light should be turned off.

Obviously a photosensitive sensor is called for here. In fact, this project is a modified combination of two other projects presented in this book—the light activated gate (Project #8), and the appliance controller (Project #11).

The complete schematic for the automatic night light project is shown in Fig. 3-7, with its parts list.

The light is simply a small dc bulb, like those used in flashlights. It only needs to be bright enough to illuminate a keyhole, or let you see where the main light switch

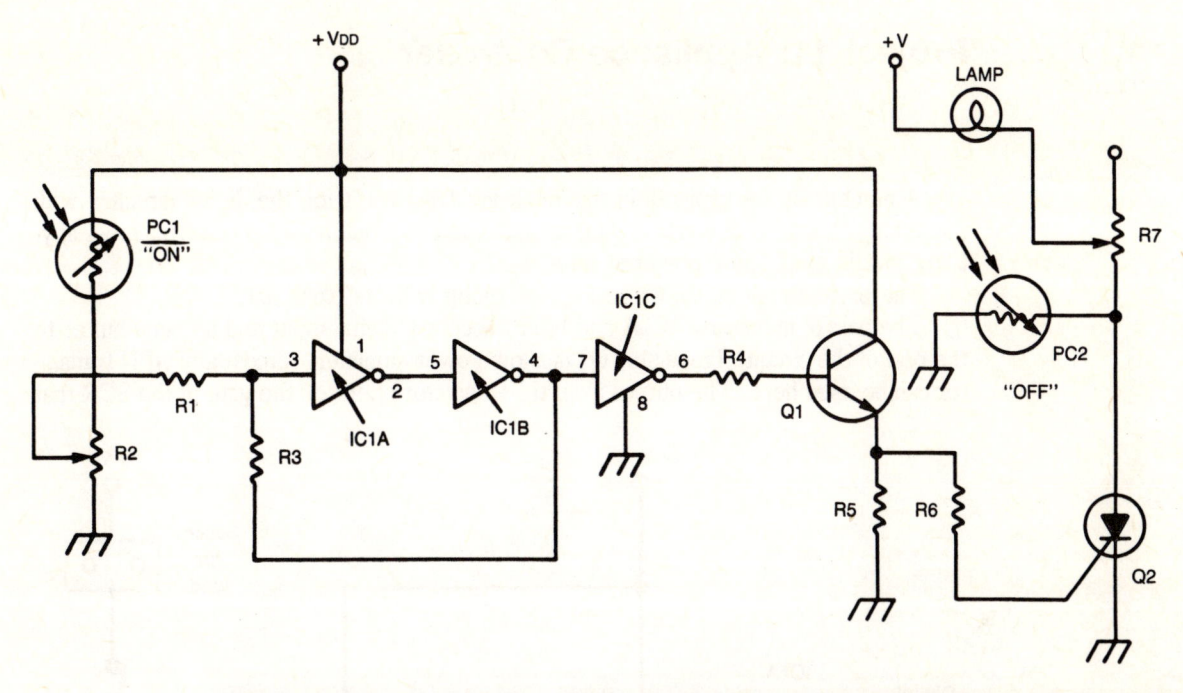

#10 PROJECT PARTS LIST

COMPONENT	DESCRIPTION
IC1	CD4049 hex inverter
Q1	NPN transistor (2N3904, 2N2222, or similar)
Q2	SCR to suit load (lamp)
PC1, PC2	photoresistor
R1	10kΩ resistor
R2, R7	100kΩ potentiometer (sensitivity adjust)
R3	100kΩ resistor
R4, R5	1kΩ resistor
R6	120Ω resistor

Fig. 3-7. This is the circuit for the automatic night light project.

is. You will probably want to devise some sort of shade to cut down the glare of the bare bulb.

The "on" sensor should be placed where it will have a clear "view" of the ambient light, and won't frequently be covered by stray shadows. The night light should be positioned so that it does not shine directly on the "off" sensor, or the circuit may not shut itself down.

Two sensor activated gates are used. One senses when the light falls below a preset level, and turns on the lamp. The other senses when the light exceeds a minimum level, and turns the lamp back off.

51

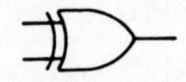

Project 11: Appliance Controller

A number of the projects in this book involve interfacing the digital circuitry with something "in the real world". This project can use a digital signal to turn almost anything that is electrically powered on and off.

The basic circuit is shown in Fig. 3-8 along with its parts list.

The pair of inverters (IC1A and IC1B) accept a digital input and act as a buffer to the rest of the circuit. Transistor Q1 is a low-power amplifier. Almost any NPN transistor can be used here. The output from the transistor operates the gate of the SCR that

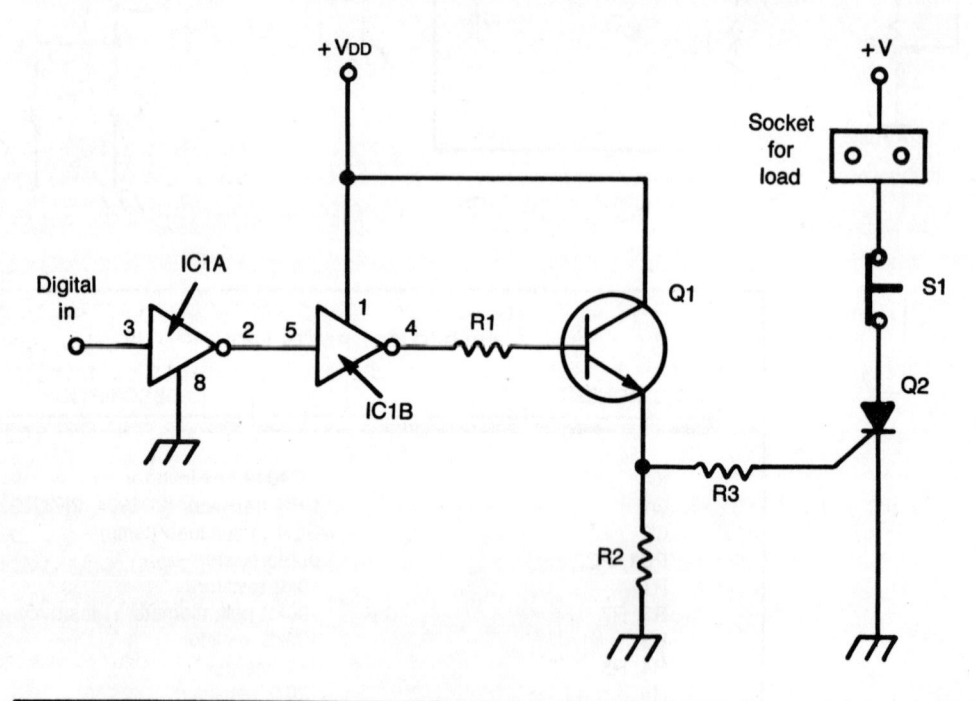

#11 PROJECT PARTS LIST

COMPONENT	DESCRIPTION
IC1	CD4049 hex inverter
Q1	NPN transistor (2N3904, 2N2222, or similar)
Q2	SCR—selected to suit intended load
R1, R2	1kΩ resistor
R3	120Ω resistor
S1	Normally Closed momentary action SPST push-switch

Fig. 3-8. A digitally activated SCR can control an external device.

is selected for the intended load. The SCR will not allow current to flow through itself (and the load) until its gate is triggered. Current can then flow until reset switch S1 is briefly opened. Obviously this switch must be a momentary action Normally Closed type.

The socket shown in the diagram can be any connection suitable for applying power to the desired load.

Notice that this circuit is for *dc powered loads only*. If you try to apply an ac voltage to the load, the SCR will cut itself off at the midpoint of each cycle. You can modify the circuit for ac loads by substituting a triac in place of the SCR. Be very careful not to allow any possibility for short circuits. If you decide on an ac powered load it would be a very good idea to use an optoisolator to eliminate the worry of an unexpected short circuit feeding ac back into the digital circuitry. The optoisolator is absolutely essential if the circuit is to be controlled by any kind of touch switch.

VARIABLE RESISTANCE/CAPACITANCE

In many electronics circuits, various parameters can be changed by altering a resistance, or a capacitance. Such changes can often be made while the circuit is in operation by using a potentiometer (variable resistor) or a variable capacitor. These components tend to be rather bulky. Variable capacitors are usually hard to find. These devices must be manually operated.

Wouldn't it be nice if there was some way to automatically vary circuit parameters while the circuit is in operation? You could set up some kind of electromechanical servo system that uses a motor and pulleys to turn the shaft of a mechanical control. This approach is inelegant, and worse, it is very tricky to set up, usually requiring frequent realignment.

Fortunately, a digital IC provides an excellent solution to the problem of automated circuit parameter changes.

The CD4066 quad bilateral switch is a very useful IC. It contains four internal switches that operate independently of each other via digital control signals. Each switch is controlled by its own bit. These switches function like true mechanical switches too. They aren't just digital gates. This means that either digital or analog signals can be switched. Only the control inputs are limited to digital signals. The switches are not polarized. Signals will pass through them in either direction. This is why they are called *bilateral*.

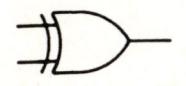

Project 12: Programmable Resistance/Capacitance

The pin-out diagram for the CD4066 quad bilateral switch IC is shown in Fig. 3-9.

Figure 3-10 shows how the CD4066 can be used as a programmable resistance. The two unconnected points, marked by triangles in the schematic, can be wired into any circuit in place of a fixed resistor.

Each switch in the CD4066 is independent of the others. This means they can be used separately and in any combination. A HIGH control signal closes the appropriate

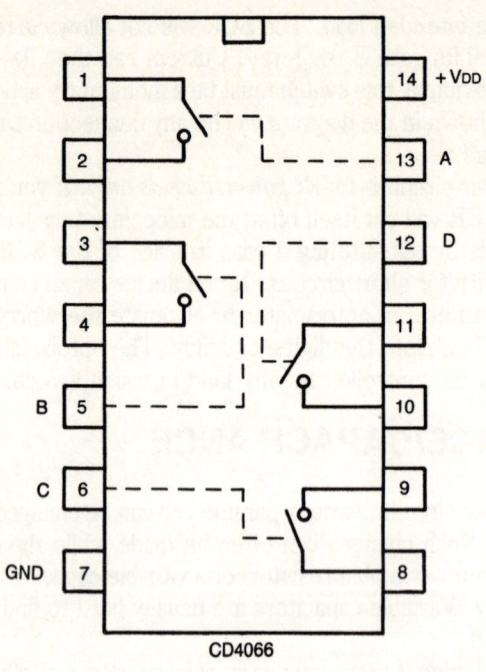

Fig. 3-9. The CD4066 contains four digitally controllable switches.

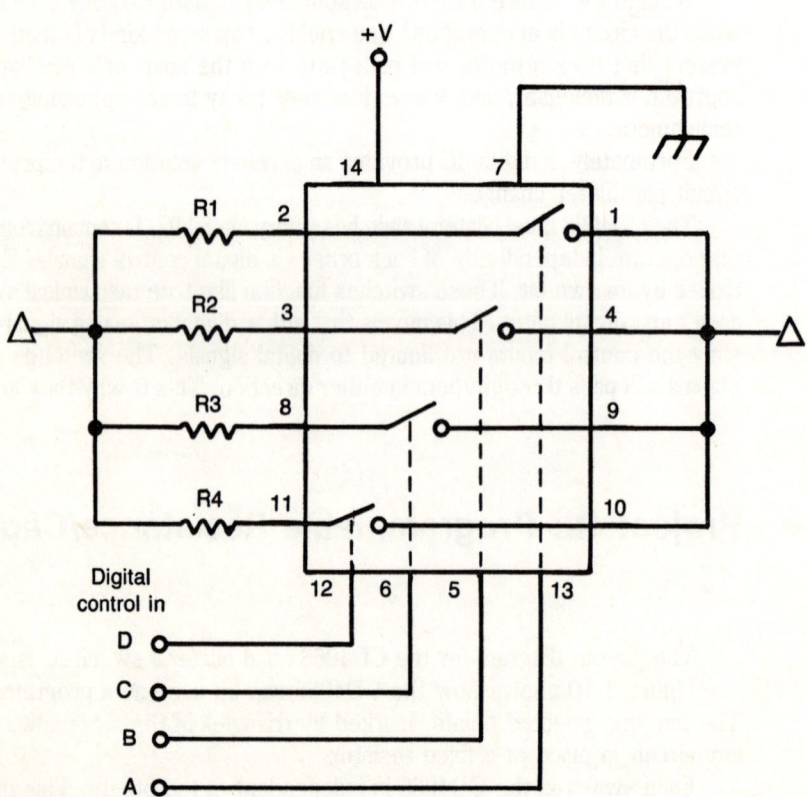

Fig. 3-10. This circuit allows you to digitally program a resistance.

54

switch, while a LOW control signal opens it. With four independent switches, there are 16 possible combinations. If all four control signals are LOW (0000) none of the resistors will be in the network. Theoretically there should be infinite resistance. Actually, there is some leakage resistance through the open switches in the CD4066, so the actual resistance under such circumstances will be very high, but not infinite.

To see just how this circuit works, we will assume the following resistance values. Of course, in your project you can use any resistor values that suit your particular application. For purposes of illustration we will use:

$$R1 = 2.2k\Omega$$
$$R2 = 1k\Omega$$
$$R3 = 470\ \Omega$$
$$R4 = 100\Omega$$

With four control bits, there are sixteen possible combinations, each producing its own unique resistance value shown in Table 3-2:

CONTROL	RESISTORS	RESISTANCE
0000	none	(infinite)
0001	R4	100 Ω
0010	R3	470 Ω
0011	R3,R4	82.5 Ω
0100	R2	1000 Ω
0101	R2,R4	91 Ω
0110	R2,R3	320 Ω
0111	R2,R3,R4	76 Ω
1000	R1	2200 Ω
1001	R1,R4	96 Ω
1010	R1,R3	387 Ω
1011	R1,R3,R4	79.5 Ω
1100	R1,R2	687.5 Ω
1101	R1,R2,R4	87 Ω
1110	R1,R2,R3	279 Ω
1111	all	73.5 Ω

Table 3-2.

When more than one resistor is selected, the resistances are in parallel. You should recall that the formula for resistances in parallel is:

$$1/Rt = 1/R1 + 1/R2 \ldots + 1/Rn$$

The total parallel resistance will always be smaller than the smallest resistor in the parallel combination.

The same basic circuitry can be used with capacitors, as illustrated in Fig. 3-11.

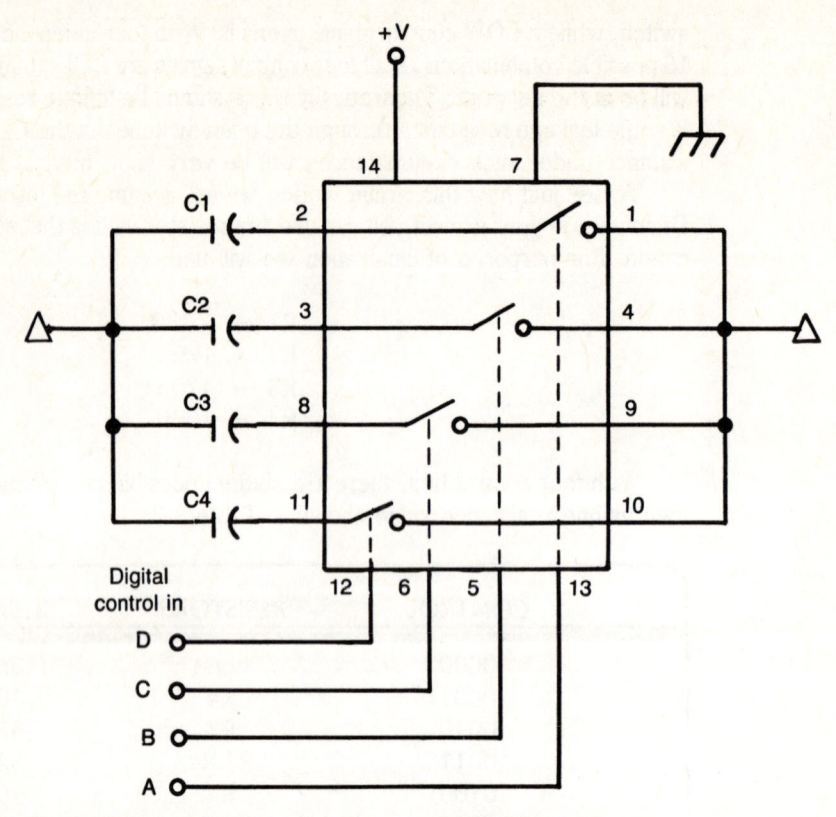

Fig. 3-11. This is the circuit for a digitally programmable capacitance.

Remember that capacitances in parallel add:

$$Ct = C1 + C2 \ldots + Cn$$

As an example, we will assume the following capacitors are being used:

$$\begin{aligned} C1 &= 0.5~\mu F \\ C2 &= 0.1~\mu F \\ C3 &= 0.05~\mu F \\ C4 &= 0.03~\mu F \end{aligned}$$

Using these component values, the combinations are possible shown in Table 3-3:
This is a very simple, but powerful and versatile project. It can be used in countless applications.

CONTROL	CAPACITORS	CAPACITANCE
0000	none	(open circuit)
0001	C4	0.03 µF
0010	C3	0.05 µF
0011	C3,C4	0.08 µF
0100	C2	0.1 µF
0101	C2,C4	0.13 µF
0110	C2,C3	0.15 µF
0111	C2,C3,C4	0.18 µF
1000	C1	0.5 µF
1001	C1,C4	0.53 µF
1010	C1,C3	0.55 µF
1011	C1,C3,C4	0.58 µF
1100	C1,C2	0.6 µF
1101	C1,C2,C4	0.63 µF
1110	C1,C2,C3	0.65 µF
1111	all	0.68 µF

Table 3-3.

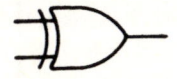

Project 13: Digital Relay Driver

In many applications we will want to control some external device with a digital signal. One way to do this is with a SCR, as in Project #11. However, a SCR will only work with dc loads, and it can be tricky turning it back off under digital control.

Another approach to digital control of external devices is to use a relay. A typical circuit for this is shown in Fig. 3-12. A suitable parts list for this project appears with it.

Actually, this circuit is fairly simple, and not unlike the SCR control circuit elsewhere in this book. The load can be either ac or dc. Since there is no direct connection between the digital control circuit and the load circuit, the relay driver can be safely used with a touch switch.

The two inverter stages simply act as a buffer so that the relay coil will not load down any other gates in the digital circuit. A logic 1 will close the relay, while a logic 0 will open it. Notice that this circuit is for momentary control only. The relay remains activated only as long as the driving input signal remains in the HIGH state. A latching relay could be used if the application demanded it. Unfortunately, latching relays tend to be more expensive, and harder to locate than the regular type.

Transistor Q1 is a simple current amplifier to boost the current level of the digital signal sufficiently to activate the relay's contacts. Almost any low power NPN transistor you happen to have handy can be used in this application.

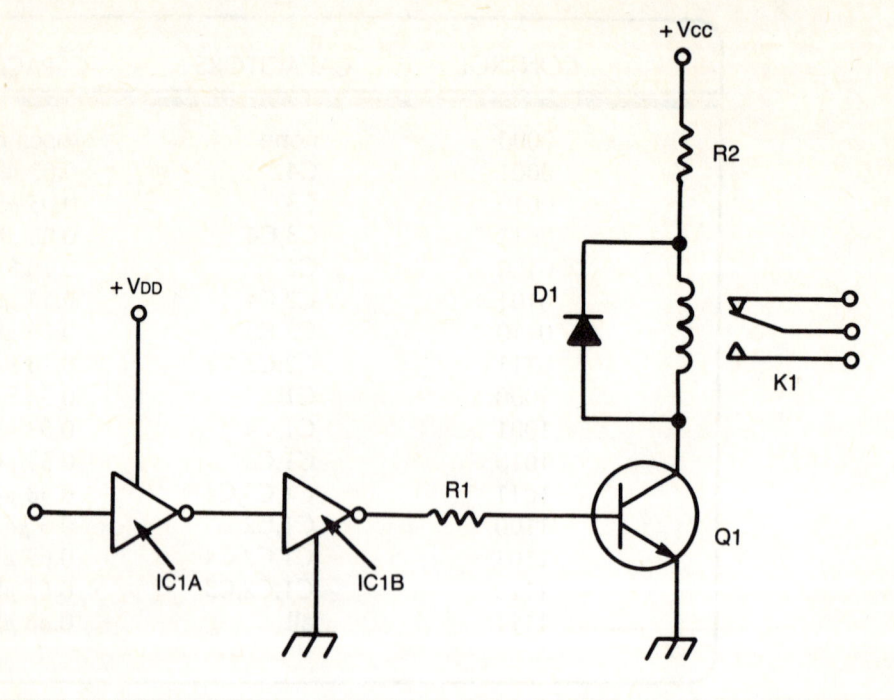

#13 PROJECT PARTS LIST

COMPONENT	DESCRIPTION
IC1	CD4049 hex inverter
Q1	NPN transistor (2N3904, 2N2222, or similar)
D1	diode (1N4148, 1N914, or similar)
R1	1kΩ resistor
R2	100Ω resistor
K1	relay appropriate to load

Fig. 3-12. This circuit activates a relay under digital control.

Diode D1 is included to prevent the relay coil from burning itself out during switching. A large inductive back-EMF will be produced whenever the coil is energized or de-energized. By placing a diode across the coil, no damage will be done by this voltage burst.

No particular specifications are given for the relay in the parts list. The relay should be selected to suit the actual load used in your application. Try to over-rate the relay a bit. For instance, if the load draws 200 mA, use a relay that can handle at least 300 mA.

The relay is shown with SPDT contacts in the diagram. Of course, you can use a relay with any number of contacts in any pattern your application happens to call for.

Like many simple circuits there are literally hundreds of potential applications for this project.

4
TEST EQUIPMENT

Nothing is perfect and infallible. And that certainly includes electronic circuits and equipment. Sooner or later you are going to have to do some troubleshooting. This chapter features several handy pieces of test equipment you can build from CMOS ICs.

In working with analog circuits, a VOM (volt-ohm-milliammeter) and an oscilloscope are the most basic tools. About 90 percent of the time, these two devices are all the technician *absolutely* needs. Other types of test equipment can certainly make the technician's life easier, but usually he can get by with just a VOM and an oscilloscope.

While the VOM and the oscilloscope are useful and occasionally essential for working with digital circuits, they leave something to be desired. These devices are not designed specifically for digital signals. They can be used, but they are often a bit awkward.

Just as there are many types of specialized test equipment for analog circuits, there are specialized purposes for digital work.

The most important type of digital test equipment is the logic probe. This is simply a device that indicates the current logic set (HIGH or LOW) at a specific point in the circuit. Either a VOM or an oscilloscope could be used for this purpose, but a logic probe is simpler and more convenient.

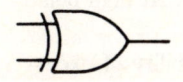

Project 14: Logic Probe

A super-simple logic probe circuit is shown in Fig. 4-1. It's hard to imagine anything simpler than this. Only two components are used—an LED, and a current dropping

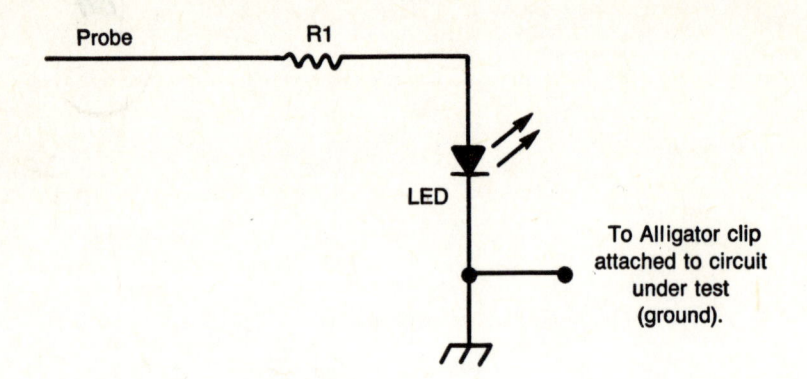

Fig. 4-1. This is a simple logic probe circuit.

resistor. An alligator clip is connected to the ground point of the circuit under test. In a manner of speaking, power for this logic probe is "stolen" from the test circuit.

The actual probe is simply a length of stiff solid wire, or a common test lead probe. If this probe is touched to a connection point (usually an IC pin) in the circuit under test, we can roughly monitor the logic state at that point. If the test point is HIGH (at logic 1), the LED will light up. If the probe is connected to anything other than a valid logic 1 condition, the LED will remain dark.

The resistor is used for current limiting. If the LED is allowed to draw too much current, it can be damaged. This resistor will generally have a value somewhere in the 200 ohm to 1000 ohm (1 kΩ) range. In most cases, a resistor with a value from 330 ohms to 470 ohms will be the best choice. The lower the resistor value, the brighter the LED will glow when lit.

This super-simple logic probe can be used without modification for any of the major logic families (CMOS, TTL, DTL, etc.). Since this circuit contains no active logic elements, and takes its power from the circuit being tested, it is nominally universal.

This circuit can easily be whipped together in just a few minutes for well under a dollar. While functional in a "quick and dirty" way, such a simple approach inevitably leaves much to be desired.

For one thing, this simple circuit does not give a definite indication of a logic 0 signal. If the LED does not light, you could have a logic 0 signal, or you might not. The probe may not be making good contact with the test point. There might even be a broken lead in the circuit so that there is no signal at all at the test point. The LED itself could be damaged. With this super-simple circuit, there is just no way of telling.

There is also no way to determine if the test point is holding a constant HIGH condition, or is pulsing at a high rate. If the test point is carrying a pulsed signal (switching back and forth between 1 and 0), the LED will blink on and off. But at pulse rates above a few Hz, the eye is unable to distinguish between the separate flashes. At high pulse rates, the LED will appear to be continuously lit.

Another potential problem area with this simple circuit is the possibility of excessively loading an IC output while it is being tested. This can occur when the gate is question is already driving close to its maximum fan-out potential.

These various problems and limitations can be corrected with a somewhat more deluxe logic probe circuit, like the one illustrated in Fig. 4-2. Here we have two indicator

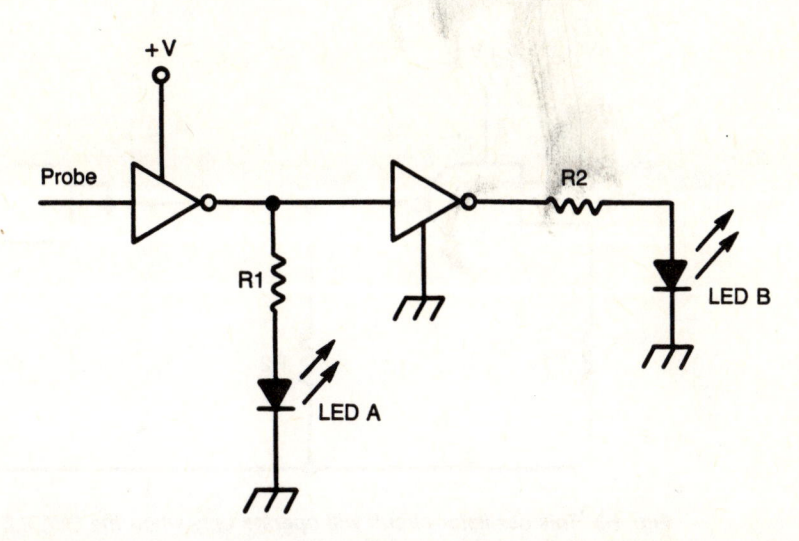

Fig. 4-2. This improved logic probe indicates both HIGH and LOW conditions.

LEDs (with their current dropping resistors), and a pair of inverters. By attaching alligator clips to the V+ and ground leads, the probe circuit can tap its power off from the circuit under test. Alternatively, the logic probe could have its own dedicated power source (probably batteries).

When a logic 0 signal is applied to the probe tip, the first inverter will change it to a logic 1, lighting LED A. The second inverter then changes the signal back to a logic 0, so LED B remains dark. Conversely, with a logic 1 signal at the probe tip, the output of the first inverter is 0, so LED A stays dark, while LED B is lit up by the logic 1 signal appearing at the output of the second inverter.

As you can see, this circuit can give a definite indication of either a logic 1 or a logic 0 state. If neither LED lights up, the probe is not making proper contact to a valid digital signal. If both LEDs appear to be lit, the probe is detecting a pulse signal. This circuit is far less ambiguous than the super-simple logic probe presented earlier. In addition, the inverter stages act as buffers, significantly reducing any loading of the circuit being tested.

We can go even further and add some additional features to our logic probe circuit to make it even more useful. Often when testing a circuit the technician has to concentrate on several things at once. For example, in testing a densely packed circuit, he might have to carefully watch *exactly* where the probe is to prevent accidental shorting, or testing the wrong pin. But how can he watch the probe tip and the LED indicators at the same time. It can be done, but it is awkward at best. In many cases, it would be very handy to have an audible indication of the signal state.

This can easily be accomplished with a gated oscillator circuit, like the one shown in Fig. 4-3. The circuit can oscillate only when a logic 1 signal is fed to the enable input. A logic 0 signal will inhibit the oscillator action. The enable signal can be taken off the inverter outputs of the basic logic probe circuit. For maximum benefit, use two gated oscillators (at widely separated frequencies). One frequency will indicate a logic 1 state, while the second frequency indicates a logic 0 at the probe input. Personally, I think

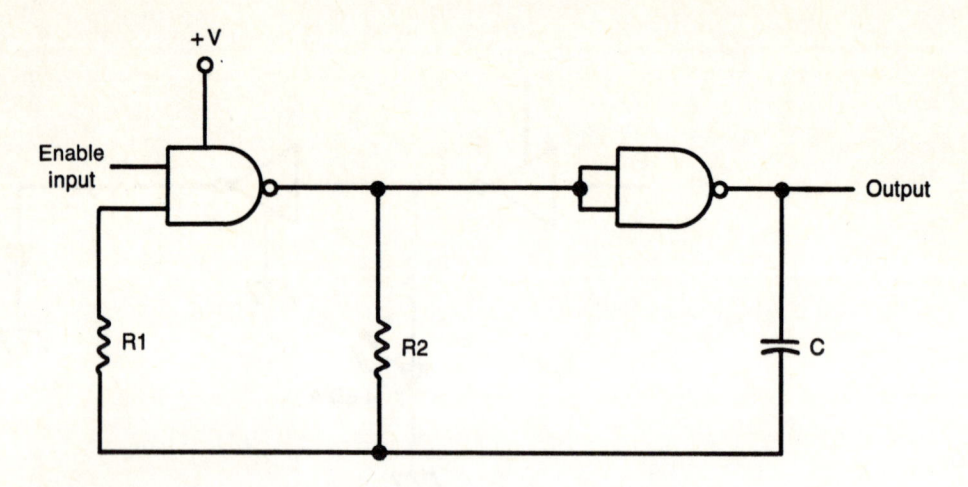

Fig. 4-3. *This oscillator circuit will operate only when the ENABLE input is HIGH.*

it makes sense to use a fairly high pitched tone for logic 1, and a low pitched tone for logic 0.

The output frequency for this circuit can be approximately found with this simple formula:

$$F = R_2C/2.2$$

Resistor R1 should have a value from about 5 to 10 times the value of R2. This resistance isn't terribly critical.

If you use two oscillators you will also be able to recognize a pulse condition. The tones will combine, producing a rather coarse buzzing quality. The exact sound will vary with the rate and duty cycle of the pulses, but it will be distinct from either of the relatively pure tones of the constant state conditions. To make pulsed signals as audible as possible, the two steady state tones should be selected so that they are not harmonically related. For example, you might use 500 Hz and 2780 Hz.

Another useful feature for a logic probe is a "pulse stretcher." Often you will need to detect a very brief single pulse. You might blink and miss the flash of the LED. Or it might even be too brief to create a visible flash. A monostable multivibrator can produce a long output pulse when it receives even a brief input pulse. The output pulse will have the same duration, regardless of the length of the input pulse. Effectively, the monostable multivibrator *stretches* the duration of the input pulse so the LED will remain lit long enough to be recognized.

Figure 4-4 shows a typical monostable multivibrator circuit. A 7555 timer is used. This is a CMOS version of the popular 555 timer. You could use a standard 555 if you prefer. A time constant of 1 to 3 seconds should be used.

A complete logic probe project with "bells and whistles" is illustrated in Fig. 4-5. A recommended parts list appears in the figure.

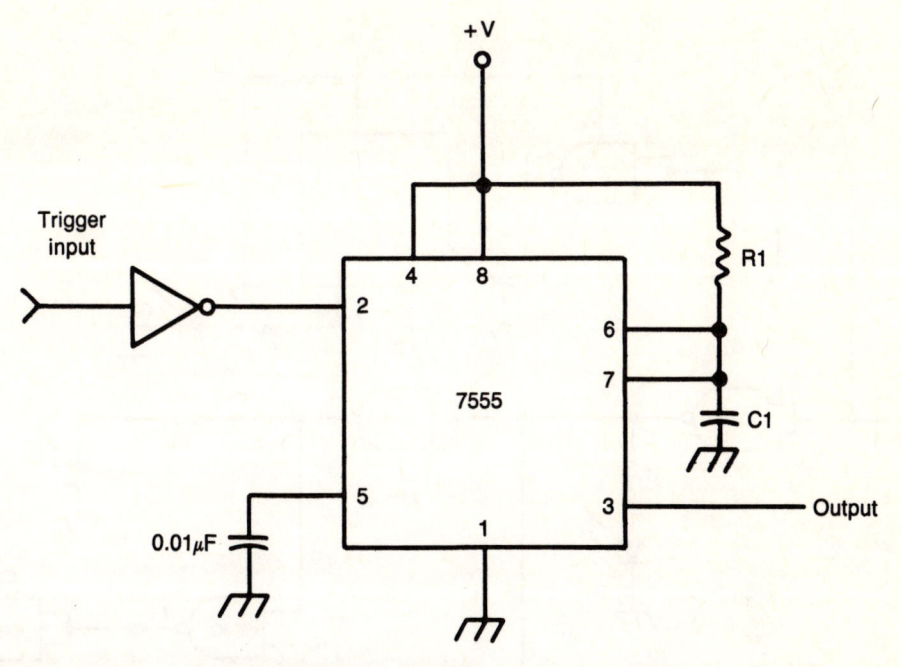

Fig. 4-4. This is a simple pulse-stretcher circuit.

#14 PROJECT PARTS LIST	
COMPONENT	DESCRIPTION
IC1, IC3	CD4011 quad NAND gate
IC2	7555 timer
D1, D2, D3	LED
C1	0.68 μF capacitor
C2	0.01 μF capacitor
C3, C4	0.1 μF capacitor
C5	50 μF electrolytic capacitor (35 volt)
R1, R2, R4	470 Ω resistor
R3	2.2 MΩ resistor
R5	68kΩ resistor
R6	10kΩ resistor
R7	390kΩ resistor
R8	56kΩ resistor
R9, R10	2.2kΩ resistor
S1	SPDT switch
SPKR	small speaker

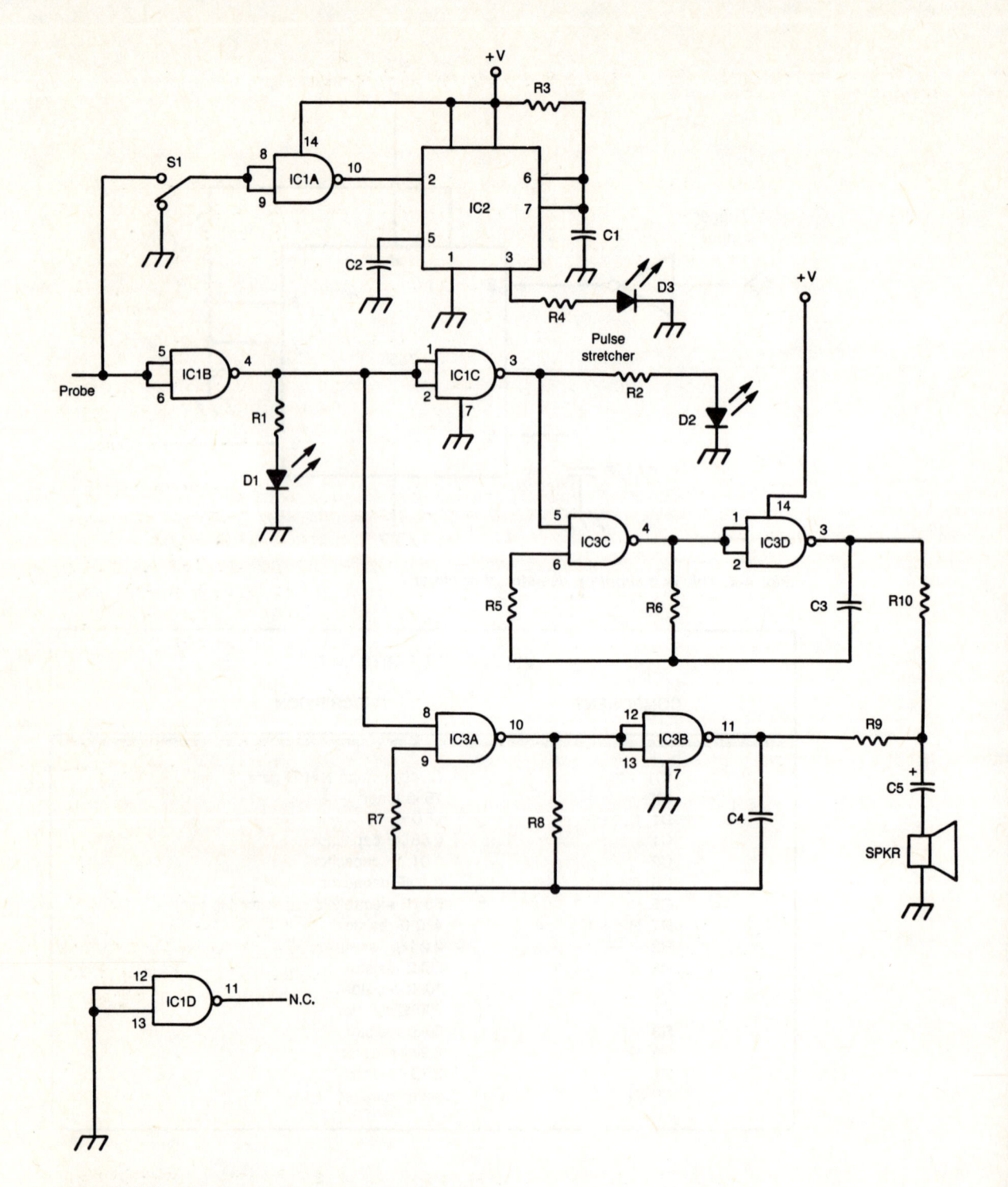

Fig. 4-5. This is the circuitry for a deluxe logic probe.

Project 15: Simple Frequency Meter

Analog frequency measurement has always been tricky and inconvenient, and usually has been plagued by touchy calibration, and inaccuracy. Digital frequency meters, on the other hand, are relatively easy to build, and can be made extremely accurate.

Even using digital ICs, a frequency meter is a fairly complex and somewhat expensive device. This project is sort of a compromise analog/digital hybrid. This simple, inexpensive circuit can be used to give a rough approximate reading of a frequency.

The schematic for this simple frequency meter project along with the parts list appears in Fig. 4-6.

The 7555 timer is wired in the astable mode and functions as a reference frequency source. To calibrate the circuit, feed a known (and as accurate as possible) frequency signal into the input, adjust potentiometer R1 until the LEDs are alternately lit and dark (that is, on-off-on-off-on, etc.).

For best (and most reliable) results, both the calibration and unknown frequency signals should be rectangle waves. If you must work with other waveshapes, it might be necessary to add a Schmitt trigger input stage, to square off the signal.

Once the circuit has been calibrated to a known frequency feed in the unknown frequency to be measured. The LEDs will light up or remain dark in some pattern. Near the center of the display one or more LEDs should be grouped together, set off by a dark LED at either end. For example, you might get a pattern like this:

on-off-on-on-off-on-on-on-off-on

Find the largest group of consecutively lit LEDs. In our example, the largest group is made up of three lit LEDs. Calling this number X, the unknown frequency can be approximately calculated with this formula:

$$F_x \cong F_r/X$$

where F_x is the unknown frequency being measured, F_r is the reference frequency used for calibration, and X is the maximum number of consecutively lit LEDs.

As an example, let's assume we had used a reference signal with a frequency of 750 Hz to calibrate the circuit. If the unknown frequency signal produces the output pattern in the preceding example, we can conclude that the unknown frequency is approximately equal to:

$$F_x \cong F_r/X$$
$$\cong 750/3$$
$$\cong 250 \text{ Hz (roughly)}$$

Obviously this isn't the most convenient device to use. It is fairly complex to interpret compared with devices that read-out directly (analog meters, true dot/bargraphs,

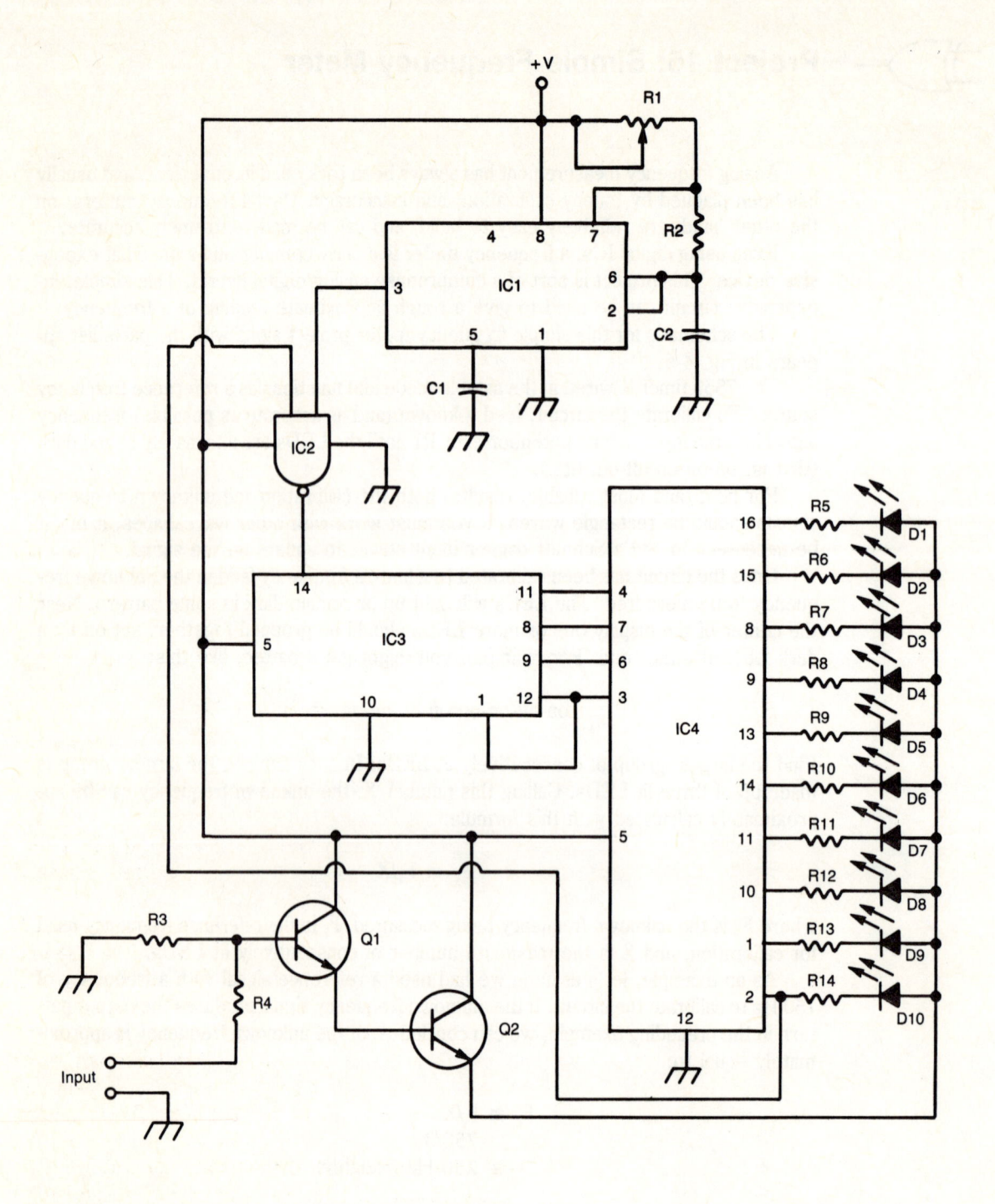

Fig. 4-6. This is the circuit for a simple frequency meter.

#15 PROJECT PARTS LIST

COMPONENT	DESCRIPTION
IC1	7555 timer
IC2	CD4011 quad NAND gate
IC3	74C90 decade counter
IC4	74C41 BCD/Decimal decoder
Q1, Q2	NPN transistor (2N2222, or similar)
D1 - D10	LED
C1	0.01 µF capacitor
C2	0.022 µF capacitor
R1	100kΩ potentiometer
R2	1kΩ resistor
R3	100kΩ resistor
R4	680kΩ resistor
R5 - R14	330Ω resistor

or numerical readout displays). Also, the resolution is quite poor. In the example we just used, the actual frequency of the unknown signal might be anywhere between 190 Hz and 375 Hz. This circuit is also phase dependent.

The pattern of lit LEDs might fluctuate wildly without providing any meaningful reading at all.

Even with all these drawbacks this circuit does offer a very inexpensive way to roughly check out frequencies from less than 1 Hz to about 50 kHz. This is also a very fascinating circuit to experiment with, which is why it is included in this book.

DIGITAL FREQUENCY METER FUNDAMENTALS

Project #15 was a simple, somewhat crude frequency measuring device. In this project we will build a true digital frequency meter. Before getting to the project itself, we should take a moment to consider how digital frequency meters in general work.

Most commercially available frequency counters use the *window* counting method. A sample of the input signal is allowed to pass through a gate. This sample lasts a specific and fixed period of time. By counting the number of pulses during this sample period, the input frequency can be reliably determined with some degree of accuracy.

A block diagram for a typical window type digital frequency meter is shown in Fig. 4-7. The input signal is first fed through an amplifier that functions as a buffer, and boosts the signal to a usable level. This amplification stage is not always included in frequency meter circuits, but such an amplifier improves the sensitivity of the instrument, permitting accurate measurement of lower level signals.

The next stage of the circuit is a Schmitt trigger that converts (almost) any input waveshape to something close enough to a rectangle wave to be reliably recognized by the digital circuitry. If only square, rectangular, or pulse waves are to be measured, the Schmitt trigger can be omitted from the circuit.

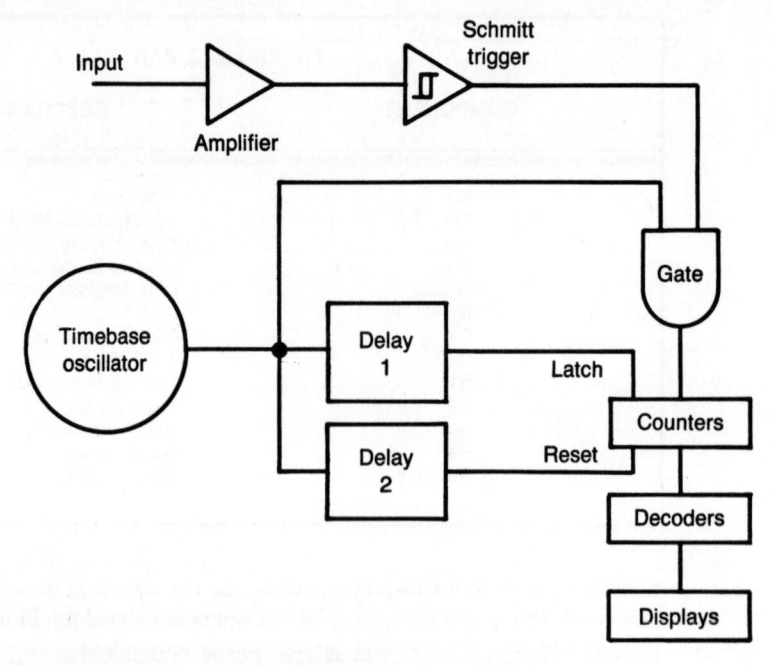

Fig. 4-7. This is a block diagram of a digital frequency counter.

The processed input signal from the Schmitt trigger is fed to one input of an AND gate. The other input to this gating stage is a regular stream of pulses from a reference oscillator, or timebase (as it is usually called for this type of application). The timebase puts out three signals (or a single signal tapped off with delay circuits, as shown in the diagram). These three timebase signals are synchronized, and their timing (*phase*) relationships are extremely critical. The three signals are illustrated in Fig. 4-8.

The first of these timebase signals (labeled GATE in the diagram) is fed to the second input of the gating circuit, effectively "opening the window" when it is at logic 1, and "closing the window" when the signal is a logic 0. The second timebase signal is delayed until after the first is over. This signal latches the outputs of the counters, so they can hold their final value long enough to produce a readable display. The third and final timebase signal resets the counters to zero in preparation of the next measurement cycle.

Incidentally, the accuracy of almost all digital frequency counters is given as X% ± 1 digit. The least significant digit may bob up and down on successive measurement cycles. This occurs whenever a partial input pulse happens to get through the window.

For a functional frequency meter, the timebase oscillator must be extremely precise in its output frequency with an absolute minimum of frequency drift. Any error here will throw off the readings, often by a significant amount. In many frequency counters, a crystal oscillator is used for the timebase.

The input frequency must be higher than the reference (timebase) frequency. If the input frequency is lower than the reference frequency, only 1 or 0 pulses could ever get through each sample window. That obviously would not result in any meaningful readings. If you need to measure lower frequencies, an additional frequency multiplier input

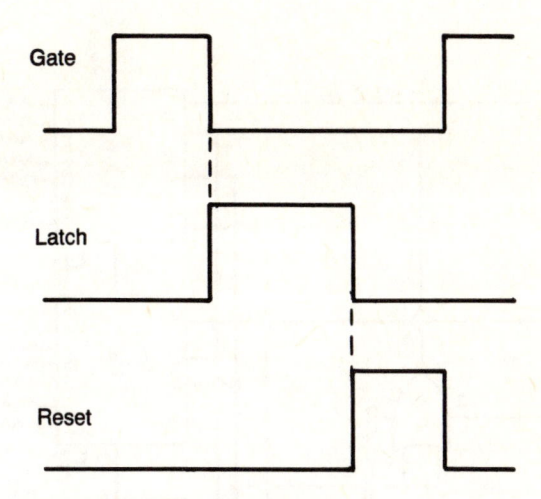

Fig. 4-8. Typical signals within a frequency counter circuit.

stage can be added between the Schmitt trigger and the gating circuit. Similarly, to measure very high frequencies that might over-range the counter stages, a frequency divider stage could be added to drop the input signal to a more convenient lower frequency.

Most commercial frequency counters have three to six counter stages for maximum counts of 999 to 999999. The project is shown here with four digits (maximum count of 9999), but you can add more counter stages if you like.

Switchable frequency multipliers and/or dividers are also often included in commercial units to allow manually selectable ranges. Decimal points may or may not be included in the display read-out. There is no reason why such modifications cannot be easily added to the project here.

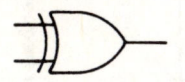

Project 16: Digital Frequency Meter

The schematic for a practical digital frequency meter project is shown in Fig. 4-9 with its parts list.

Each digit of the output display is driven by an individual CD4026 decade counter (IC2 through IC5). Four digits are shown in the diagram, but it is a very simple matter to extend the display to allow additional digits. Simply connect pin #5 from the last stage to pin #1 of the following stage. The other pins of each additional counter IC are connected in exactly the same way as shown for IC2 through IC5.

Q1 and IC1 pre-condition the signal so that it will have an acceptable level and waveshape to be reliably counted by the digital circuits.

IC6 serves as the reference oscillator, or timebase. The timebase frequency can be adjusted with R33, a 1 Megohm trimpot. Calibration of this circuit is done simply

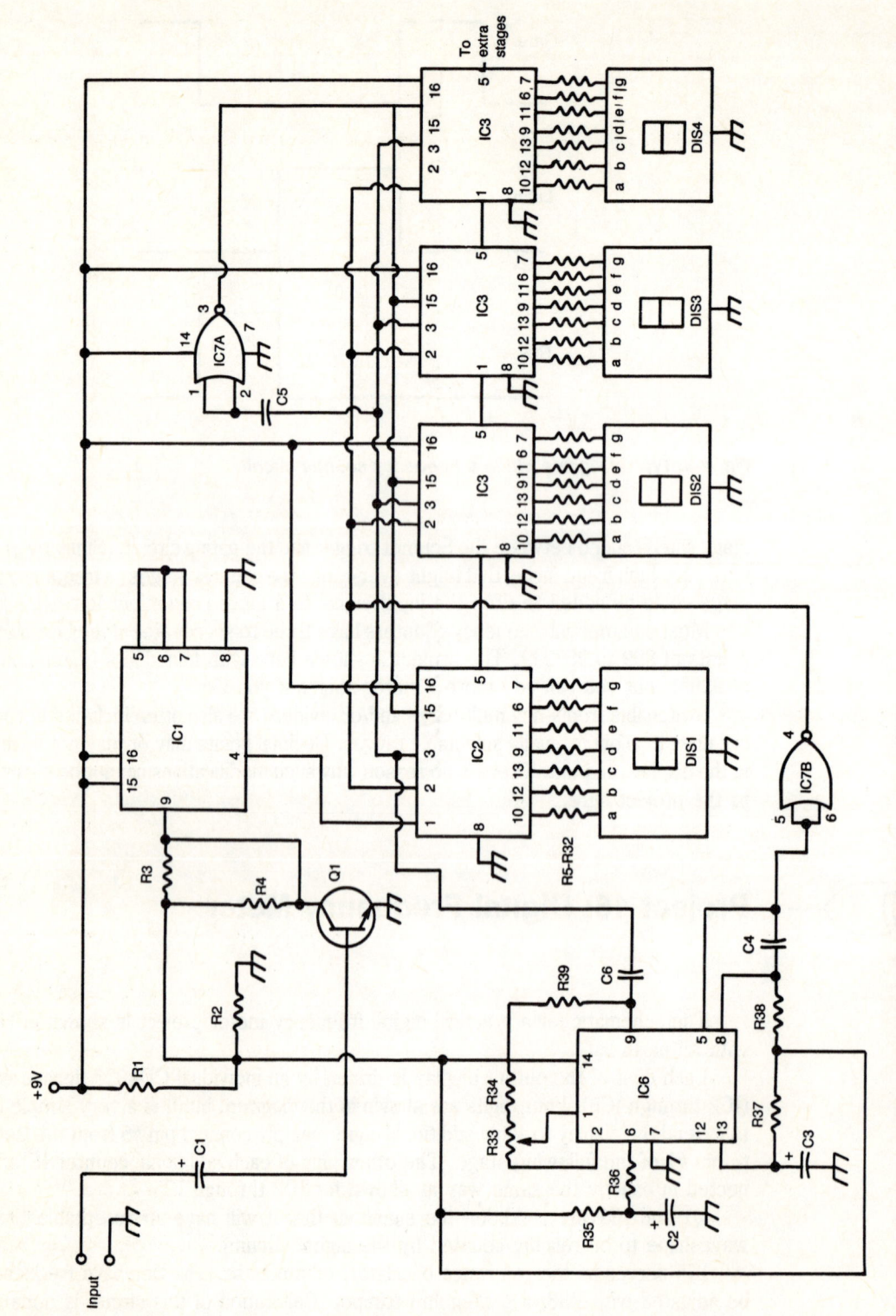

Fig. 4-9. The full schematic for the digital frequency counter project.

#16 PROJECT PARTS LIST	
COMPONENT	DESCRIPTION
IC1	14583 Schmitt trigger
IC2 - IC5	CD4026 decade counter
IC6	556 dual timer
IC7	CD4011 quad NAND gate
Q1	NPN transistor (2N3302, 2N5826, or similar)
DIS1 - DIS4	seven-segment LED display, common cathode
C1, C3	1 µF 35V electrolytic capacitor
C2	10 µF 35V electrolytic capacitor
C4, C5, C6	0.001 µF capacitor
R1	22kΩ resistor
R2	18kΩ resistor
R3, R39	100kΩ resistor
R4	10 MΩ resistor
R5 - R32	330 Ω resistor
R33	1 MΩ trimpot
R34	470kΩ resistor
R35 - R38	10kΩ resistor

by monitoring a known frequency source while adjusting R33 to obtain the correct value on the readouts.

CAPACITANCE METER CIRCUITS

Before the use of digital ICs, a number of analog capacitance meter circuits were designed, but they tended to be complex, expensive, and not particularly reliable or accurate. Digital circuitry places efficient measurement of capacitance within the reach of the average hobbyist without an exorbitant cost.

Capacitance values can be easily measured over a wide range with the use of digital ICs in a manner not dissimilar to digital voltmeters.

A block diagram of a basic digital capacitance meter circuit is shown in Fig. 4-10.

The first stage is a simple monostable multivibrator, or *one-shot*. You should recall that a monostable multivibrator has one stable state. Its output holds that stable state indefinitely, until a trigger pulse is received at the input. The output then jumps to the opposite (non-stable) state for a fixed period of time (determined by the component values in the monostable circuit). When the monostable time runs out, its output reverts to the original stable state. For our discussion we will assume that the stable state is at logic 0.

The output of the monostable multivibrator stage remains at logic 0 until the circuit is triggered. At that instant, the output of this stage switches to a logic 1 for a specific period of time, which is defined by a resistor/capacitor combination. In this application the timing resistor is a fixed value. (In some actual circuits, different resistors can be switch selectable for various ranges, but in use the resistance value actually will be non-variable.) The timing capacitor is the unknown capacitance being measured. Since we

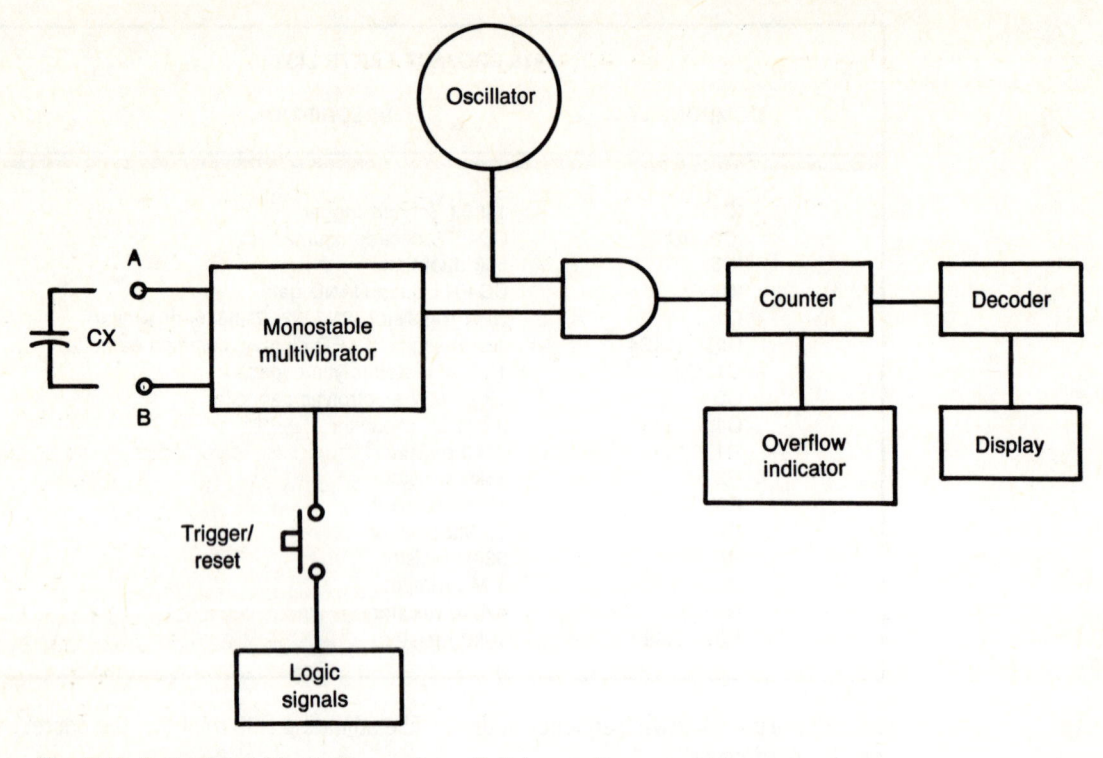

Fig. 4-10. A simplified block diagram for a digital capacitance meter.

know the resistance value, and it doesn't change, the output of the monostable multivibrator will go logic 1 when triggered for a period of time that is directly proportional to the input capacitance.

The remainder of the circuit is really just a digital voltmeter (perhaps with some minor modifications).

The output of the monostable multivibrator is one of the two inputs to an AND gate. The other input to this gate is a continuous stream of evenly spaced pulses from a square-wave oscillator. When the monostable multivibrator's output is at 0, the oscillator pulses will be blocked by the AND gate. Remember that whenever one of the inputs of an AND gate is LOW, the output will be LOW, regardless of the state at any other inputs. On the other hand, while the monostable multivibrator is putting out a logic 1, the pulse stream can pass through the AND gate to the next stage of the circuit.

The next stage is a counter that determines how many pulses manage to get through the AND gate before the monostable multivibrator time period ends. The count is checked for possible over-range, then it is decoded and the appropriate value is displayed.

Since the monostable multivibrator's HIGH output time is directly proportional to the value of the unknown capacitance (C_x), the count displayed will also be proportional to the value of C_x. By selecting an appropriate oscillator frequency, the capacitance value can be read out directly.

A push-button switch to reset the counter, and manually trigger the monostable multivibrator is usually mounted on the front panel of the instrument.

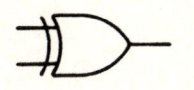

Project 17: Capacitance Meter

A practical digital capacitance meter circuit is shown in Fig. 4-11 with its parts list.

Since the voltage across the two test points (and therefore through C_x) is less than two volts, the measurement process is safe for virtually any component you might want to test.

This unit is capable of measuring capacitances from less than 100 pF to well over 1000 μF. The vast majority of capacitors a hobbyist is likely to work with will be within this range.

If electrolytic capacitors (or other polarized components) are to be tested with this circuit, be sure to hook the meter up with the correct polarity. Point A should be attached to the capacitor's positive lead. Point B is the negative connection point (ground).

Resistor R1 sets the full scale reading for the meter. This component should be a trimpot that is set carefully during calibration with a high-grade (*precision*) capacitor of a known value. Once the desired resistance has been set, the trimpot should be left alone. It should be mounted in a relatively inaccessible position. You might want to apply a dab of paint to prevent the trimpot from changing position. Even better, once the circuit has been calibrated carefully remove the trimpot from the circuit without letting its setting change at all. Measure the resistance and replace the trimpot with an appropriately valued precision resistor (1% tolerance, or better).

The value of R1 also determines the measurement range. Smaller resistance values should be used to measure larger capacitances. A 100 kΩ trimpot would be a good starting point for a $\times$1 scale, while a $\times$10 scale would be better served by a 5 kΩ trimpot. For maximum versatility, duplicate R1 for several over-lapping ranges, and use a rotary switch to select the appropriate resistor for each measurement.

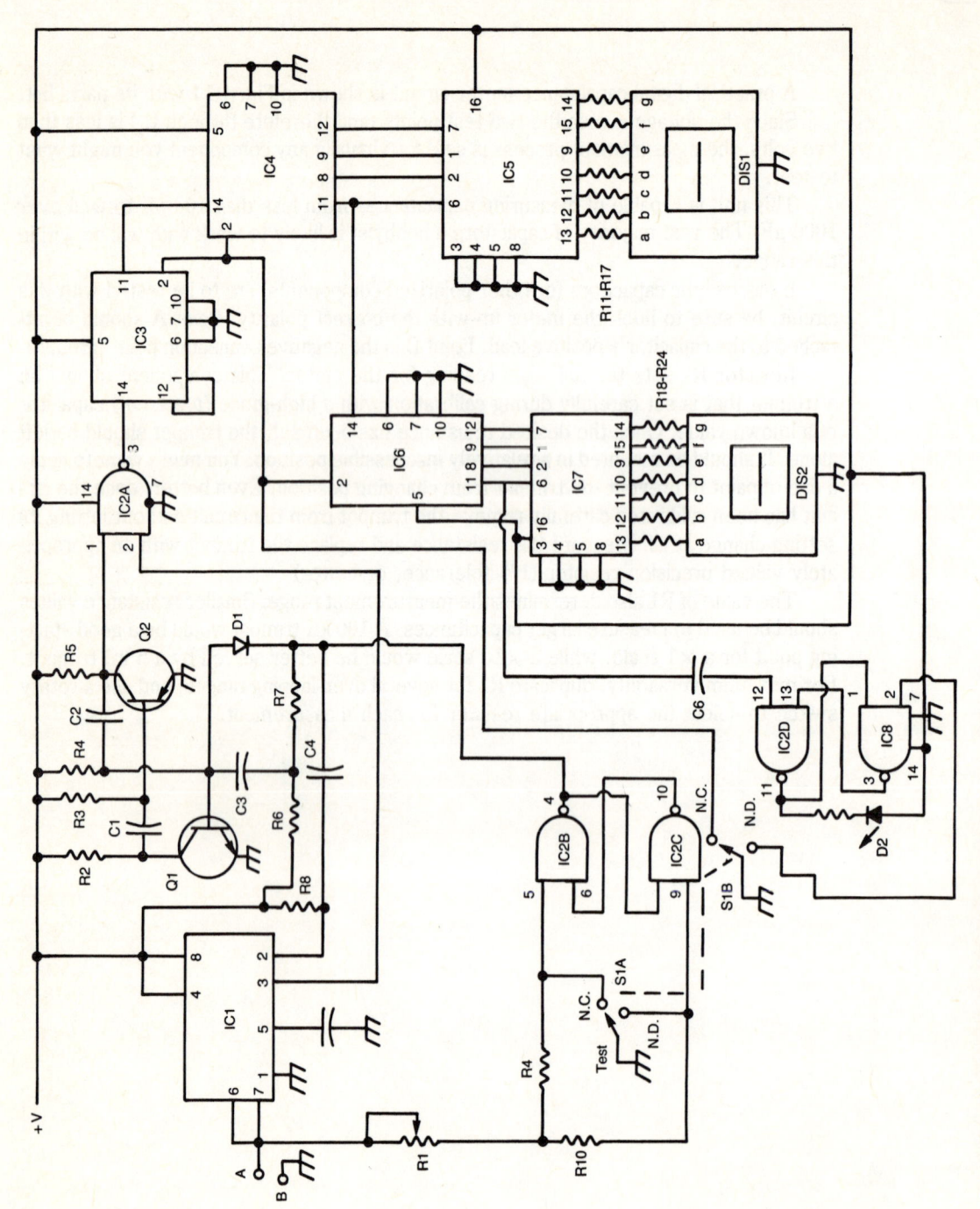

Fig. 4-11. The complete schematic for the digital capacitance meter.

74

#17 PROJECT PARTS LIST

COMPONENT	DESCRIPTION
IC1	7555 timer
IC2, IC8	CD4011 quad NAND gate
IC3, IC4, IC6	74C90 decade counter
IC5, IC7	CD4511 BCD to 7-segment decoder
Q1, Q2	NPN transistor (2N3907, or similar)
D1	1N4734 diode (or similar)
D2	red LED (overflow indicator)
DIS1, DIS2	Seven-segment LED display (common cathode)
C1, C2	0.047 μF capacitor
C3	0.1 μF capacitor
C4, C5	0.01 μF capacitor
C6	0.0022 μF capacitor
R1	calibration trimpot
R2, R5	2.7kΩ resistor
R3, R4, R7	15kΩ resistor
R6, R8	10kΩ resistor
R9, R10	1.8kΩ resistor
R11 - R24	330Ω resistor
S1	DPDT push-button switch (Normally open) push to clear and test

5
LED FLASHERS

The projects in this chapter might be considered rather frivolous. Well, so what? Who says technology can't be fun.

There is something very fascinating and even a little hypnotic about LED flasher projects. Every electronics hobbyist should try his hand at a few. You might be surprised at how easy it is to get hooked.

LED flashers aren't really completely useless. They are very eye-catching, so they can be used in warning indicators, and advertising or other displays. They can also be used in toys, and they're great for decorative purposes.

For some reason, one of the most popular types of projects has always been the light flasher or blinker. These projects are especially popular among beginners, because they tend to be quite easy to build. But even experienced hobbyists turn to such projects occasionally, just for kicks.

The popularity of this type of project might seem a bit odd, since these circuits don't really do anything practical. They simply turn one or more light sources on and off in a regular pattern. The fact is, light flashers are fun, and, if you insist on practicality, they can be used as an eye-catching display device or warning indicator.

Before the digital revolution, most light flasher circuits were built around flashlight bulbs or neon lamps. Today such circuits generally use LEDs. LEDs offer several advantages. They don't tend to burn out, they use less power, they're smaller, and frankly, they look a lot snazzier.

Project 18: LED Flasher

A simple LED flasher circuit built around two CMOS NOR gates is shown in Fig. 5-1. The blinking rate is controlled primarily by resistor R2, and capacitor C1. Experienced hobbyists should recognize this as a simple low-frequency oscillator (clock) circuit.

Reducing the value of either R2 or C1 (or both) will cause the LED to blink on and off at a faster rate. If the flashing rate is made too fast (more than about 6 to 10 times a second), the LED will appear to be continuously lit, because the human eye cannot react fast enough to distinguish the individual blinks. Of course, increasing the value of either or both of the frequency determining components (R2 and C1) will slow down the flash rate.

For a variable flash rate, you could use a small value potentiometer in series with a large fixed resistor in place of R2. The value of R1 should be 5 to 10 times the value of R2.

Resistor R3 is used for current limiting to protect the LED. Changing the value of this resistor will change the brightness of the LED when it is on. The smaller the resistance, the brighter the LED will glow. This resistor should have a value in the 200 ohm to 1000 ohm (1kΩ) range.

By adding a third NOR gate stage, as shown in Fig. 5-2, we can convert the circuit into an alternate LED flasher. The LEDs will blink in turn. When LED1 is on, LED2

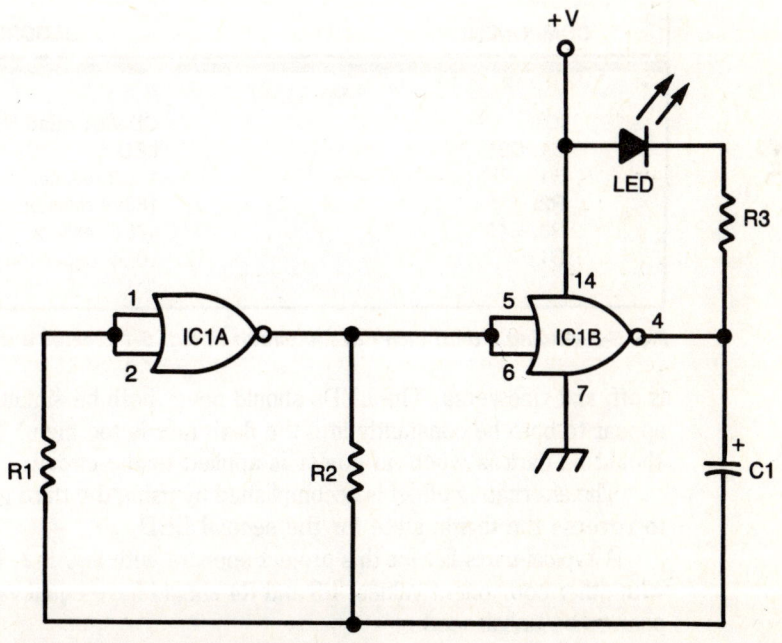

Fig. 5-1. This is a simple LED flasher circuit.

Eliminate IC1C, D2, and R4 for one LED version.

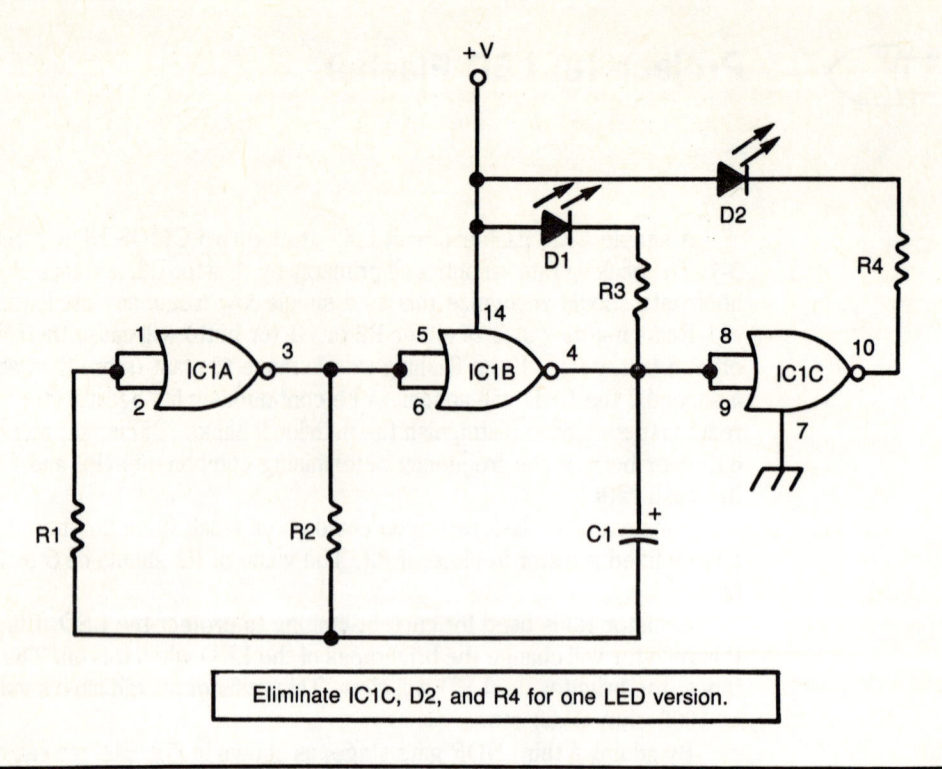

#18 PROJECT PARTS LIST

COMPONENT	DESCRIPTION
IC1	CD4001 quad NOR
D1, D2*	LED
R1	1 MΩ resistor
R2	180kΩ resistor
R3, R4*	470Ω resistor
C1	10 μF capacitor—electrolytic (35 volt)

Fig. 5-2. Adding a third stage to the circuit of Fig. 5-1 creates a dual LED flasher.

is off, and vice versa. The LEDs should never both be simultaneously lit. (They might appear to both be constantly lit if the flash rate is too high.) The only time both LEDs should be dark is when no power is applied to the circuit.

The alternating effect is accomplished by using the third gate section as an inverter to reverse the digital state for the second LED.

A typical parts list for this project appears with Fig. 5-2. By all means, experiment with other component values, R3 and R4 should have equal values or the LEDs will appear to be unbalanced.

For the component values listed the flash rate will be about 1 Hz, or approximately one flash per second.

LED flashers are popular and interesting, but they quickly become rather boring because of their inherently redundant pattern. The project described in this section will flash sixteen LEDs on and off, one at a time, in an apparently random pattern. It is hard to predict which LED will be lit next. The heart of this circuit is the 74C193 up/down counter.

Most counters only count in an upward direction from zero to a higher value. For instance, a normal four stage binary counter will exhibit the output pattern shown in Table 5-1:

DECIMAL	BINARY	DECIMAL	BINARY
0	0000	8	1000
1	0001	9	1001
2	0010	10	1010
3	0011	11	1011
4	0100	12	1100
5	0101	13	1101
6	0110	14	1110
7	0111	15	1111

Table 5-1.

In some applications it might be desirable to have a counter start at the maximum count value and work its way back down to zero, like in Table 5-2:

DECIMAL	BINARY	DECIMAL	BINARY
15	1111	7	0111
14	1110	6	0110
13	1101	5	0101
12	1100	4	0100
11	1011	3	0011
10	1010	2	0010
9	1001	1	0001
8	1000	0	0000

Table 5-2.

In other words, the counting sequence is generated backwards. This is not too difficult to achieve. We can just invert each of the binary outputs. Alternatively, if the counter is built up from independent flip-flop stages, the $\overline{Q}$ (NOT Q) outputs can be used instead of the regular Q outputs. In either case, the output will be the opposite of its ordinary state, producing a count-down sequence.

But what if the count direction must be reversed in operation? It can be done with a complex gating network, but IC designers have identified this potential problem and have produced a dedicated solution. The 74C193, which is shown in Fig. 5-3 is a dedi-

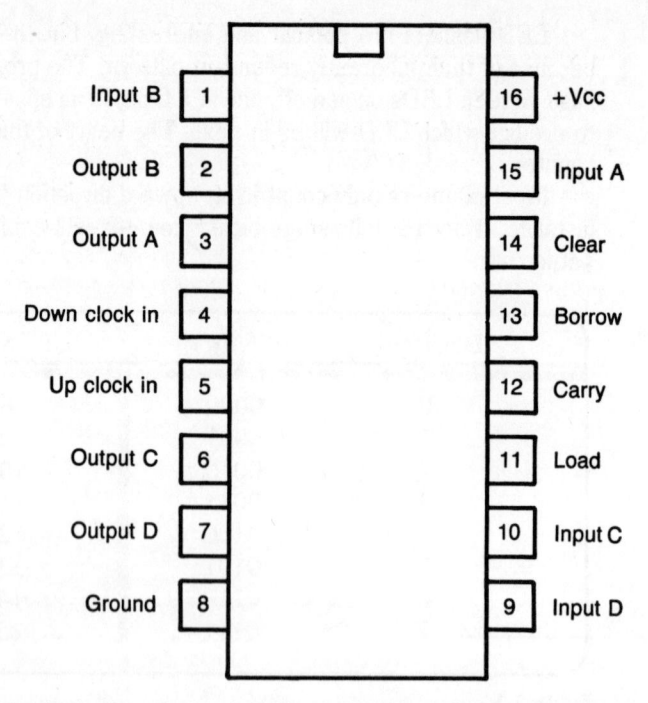

```
                    ┌──┐
    Input B    1 ─┤16├─ +Vcc
    Output B   2 ─┤15├─ Input A
    Output A   3 ─┤14├─ Clear
    Down clock in 4 ─┤13├─ Borrow
    Up clock in   5 ─┤12├─ Carry
    Output C   6 ─┤11├─ Load
    Output D   7 ─┤10├─ Input C
    Ground     8 ─┤ 9├─ Input D
```

Fig. 5-3. The pseudo-random flasher is built around the 74C193 up/down counter.

cated counter chip that can count in either direction (up or down), depending on the logic signals fed to pins #4 and #5.

Applying a logic 1 to the clear pin (#14) forces the outputs to 0000, regardless of the previous count value. The counter can also be preset to any value set at inputs A through D:

INPUT	PIN #
A	15
B	1
C	10
D	9

The presetting is accomplished by temporarily setting the load input pin (#11) to logic 0.

If pin #14 (clear) is grounded (logic 0), and pin #11 (load) is connected to the V+ line (through a resistor if appropriate) (logic 1), the 74C193 will behave just like a standard four-bit binary counter, with the direction of count determined by the logic signals at pins #4 and #5.

Pins #4 and #5 are used as clock inputs. The count is incremented by one each time a logic 1 pulse is fed to pin #5. Similarly, the count is decremented one for each HIGH pulse fed to pin #4. Notice the count can reverse direction at any point. Clearly the 74C193 is an extremely versatile device. It is also available in most of the major TTL subfamilies.

By now you are probably wondering what all this has to do with a pseudo-random flasher. Now that we understand how the 74C193 works, there is no need to wait any longer.

To achieve a pseudo-random output pattern, we simply apply two different clock frequencies to pins #4 and #5. The most dramatic and random-seeming effects are created when the two input frequencies are at least moderately far from each other and do not bear any harmonic (multiple) relationship to one another.

You could use the binary outputs to drive four LEDs directly, but it's more interesting to feed the binary outputs of the counter through a demultiplexer (such as the 74C154). This gives us a one out of sixteen random flasher. Only one of sixteen LEDs will be lit at any given instant.

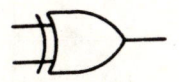

Project 19: Pseudo-Random Flasher

The circuit for the pseudo-random flasher project is shown in Fig. 5-4 accompanied by its parts list. As you can see, not much is required.

For the best results, use two separate clock sources as the input, adjusted to non-related frequencies. Alternatively, a single clock signal can be split into two with some gates. A typical approach using a single quad NAND gate IC (CD4011) is illustrated in Fig. 5-5. To maximize the pseudo-random effect, two of the demultiplexer's outputs are fed back to combine with the incoming clock signal. Pins #3 and #9 are indicated in the diagram, but you can use any of the 74C154's output pins.

This basic idea can be expanded as much as desired by adding more gates and/or feedback paths. The gating network shown in Fig. 5-6 uses a quad NAND IC (CD4011) and a quad NOR IC (CD4001). This is fascinating to experiment with on a solderless breadboard. Try as many different inputs as you can think of.

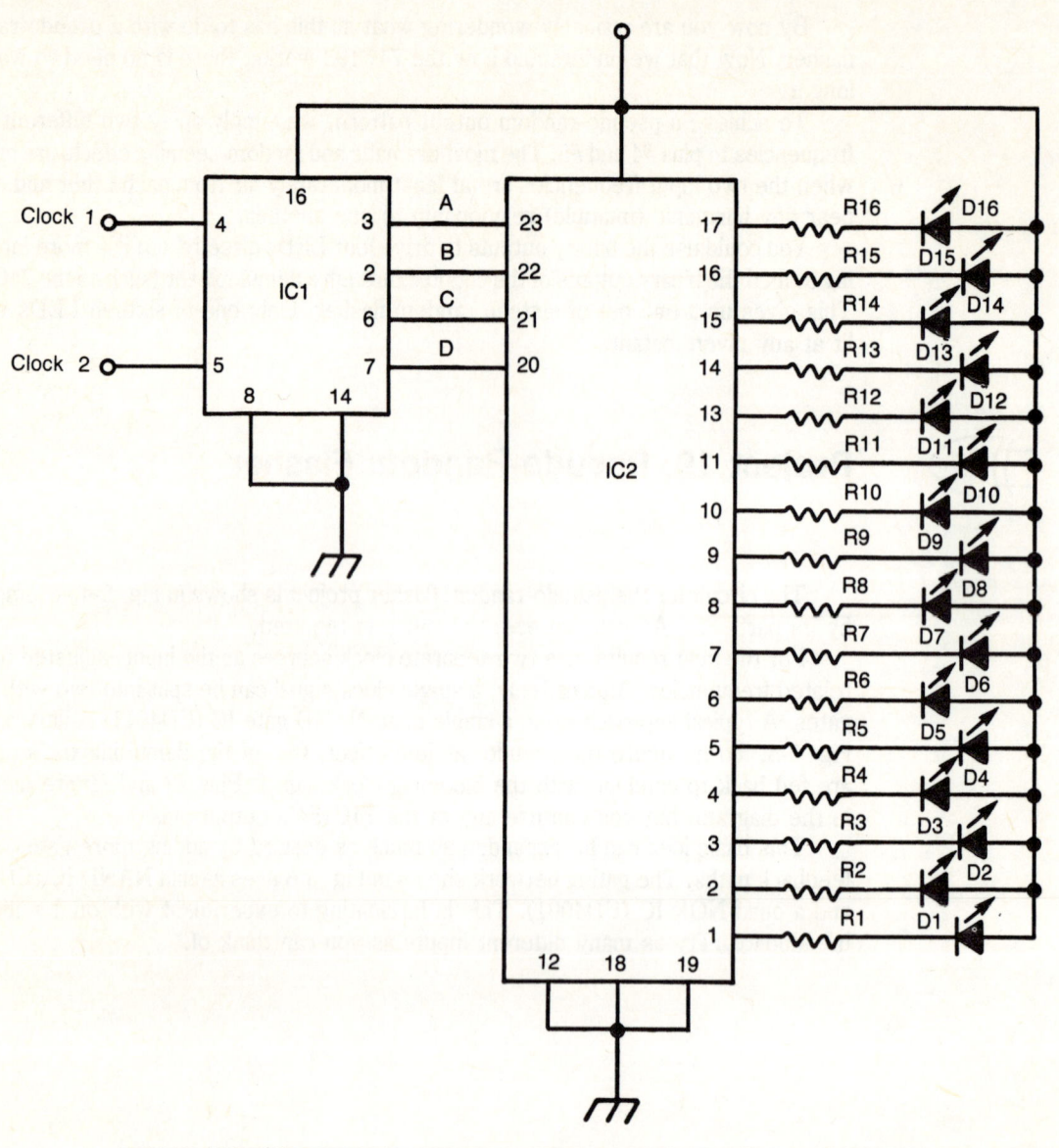

#19 PROJECT PARTS LIST

COMPONENT	DESCRIPTION
IC1	74C193
IC2	74C154
D1 - D16	LED
R1 - R16	470Ω resistor

Fig. 5-4. This circuit will flash LEDs on and off in apparently random sequence.

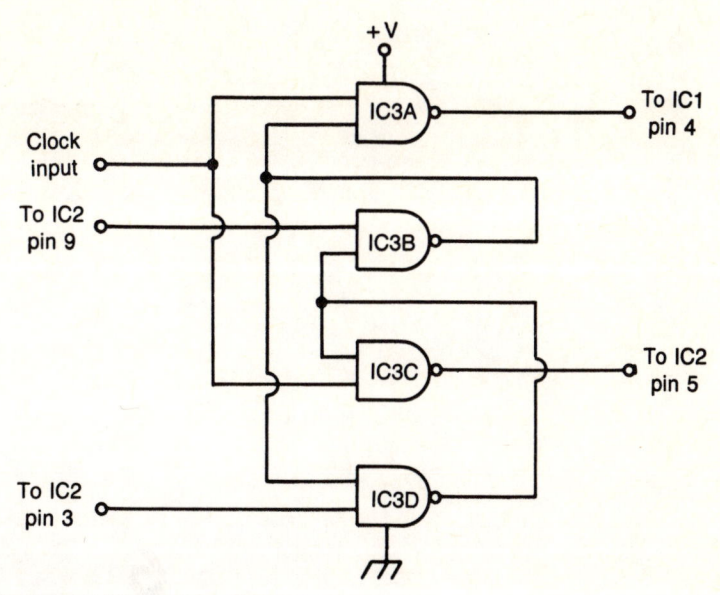

Fig. 5-5. A single CLOCK can be split into two.

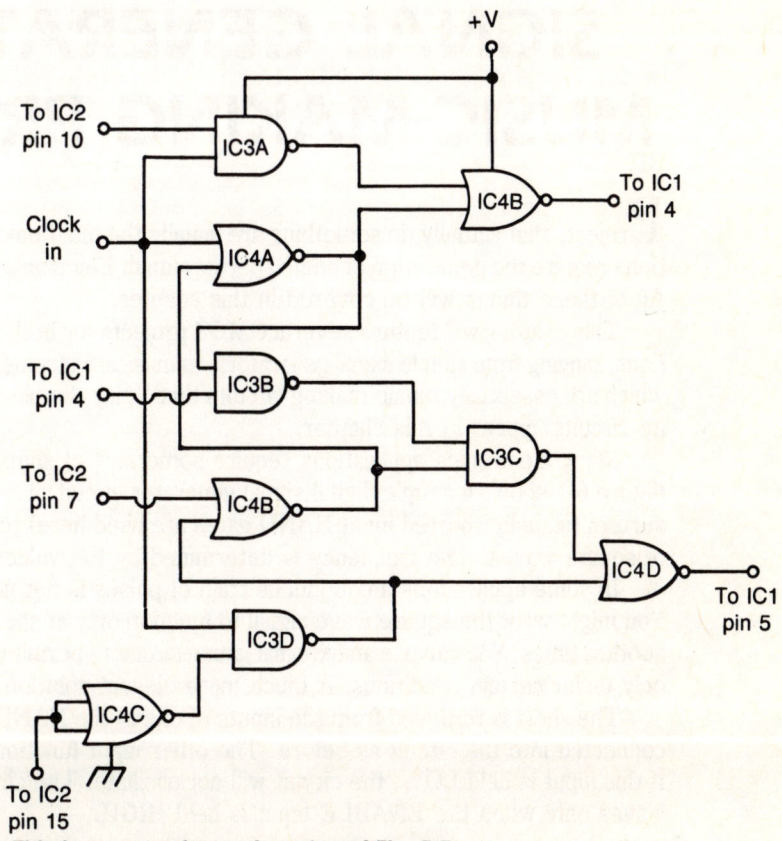

Fig. 5-6. This is a more advanced version of Fig. 5-5.

6

SIGNAL GENERATOR & MUSIC-MAKING PROJECTS

Projects that actually do something are usually the most enjoyable. Countless applications require the generation of some specific signal. Electronic music is a popular hobby. All of these things will be covered in this chapter.

This chapter will feature several CMOS projects for both analog and digital applications, ranging from simple signal generators, to musical instruments, and even sequencers, which are essentially music-making circuits that "play themselves". Some of my favorite circuits appear in this chapter.

Many electronics applications require some sort of signal generator or oscillator. Figure 6-1 shows a simple digital signal generator circuit. This circuit uses a pair of inverters (actually shorted input NAND gates are used here) to produce a regular string of square waves. The frequency is determined by the values of R3 and C1.

In some applications a continuous train of pulses is not needed or even desirable. You might want the square wave signal to function only at specific times, and disappear at other times. You can use an external gating circuit to permit the pulses to pass through only under certain conditions. A much more elegant solution is illustrated in Fig. 6-2.

The short is removed from the inputs of one of the NAND gates. One input is now connected into the circuit as before. The other input functions as an ENABLE input. If this input is held LOW, the circuit will not oscillate. The circuit will generate square waves only when the ENABLE input is held HIGH.

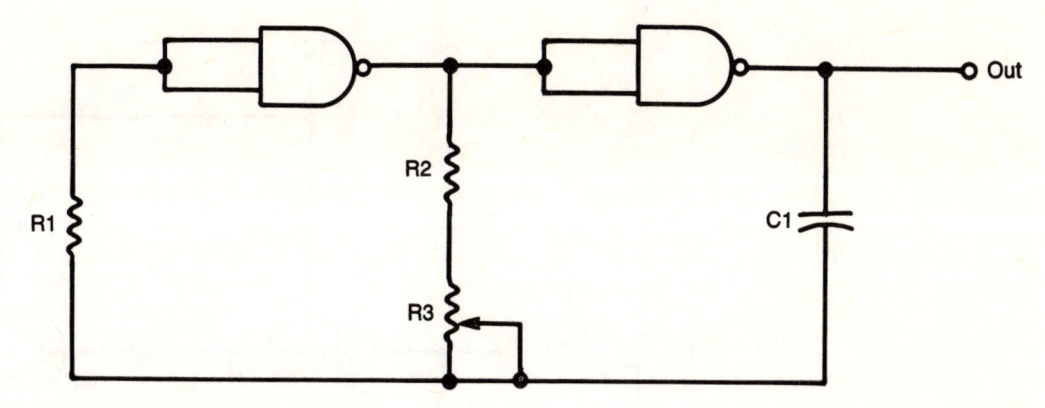

Fig. 6-1. A simple digital oscillator.

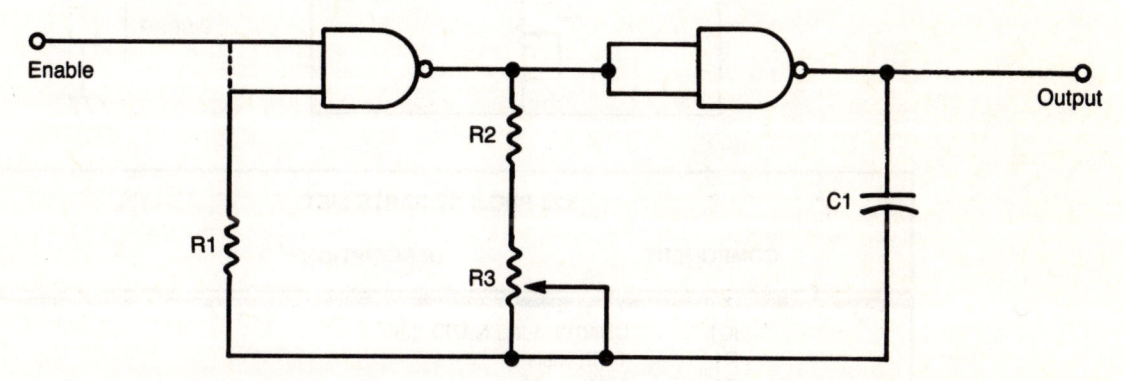

Fig. 6-2. This gated oscillator generates an output signal only when the ENABLE input is HIGH.

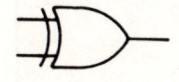

Project 20: Gated Oscillator

Figure 6-3 shows a practical application for this idea, the additional components allow the oscillator to drive a small loudspeaker. The speaker will emit a tone when the ENABLE input is HIGH. When the ENABLE input is LOW, the speaker will be silent.

Be sure to experiment with different settings of potentiometer R3. Also try substituting different values for capacitor C1. Both of these components control the frequency of the tone generated by the circuit.

Potentiometer R5 is a volume control. If you prefer, R5 and R6 can be combined into a single fixed resistor, with a value of at least 220 ohms. (R6 is included in the circuit to prevent R5 from being set at too low a resistance, which could damage transistor Q1.

Almost any NPN transistor can be substituted for transistor Q1. The types listed in the parts list are commonly available examples. A complete suggested parts list is included with Fig. 6-3.

Fig. 6-3. This is a gated tone generator circuit.

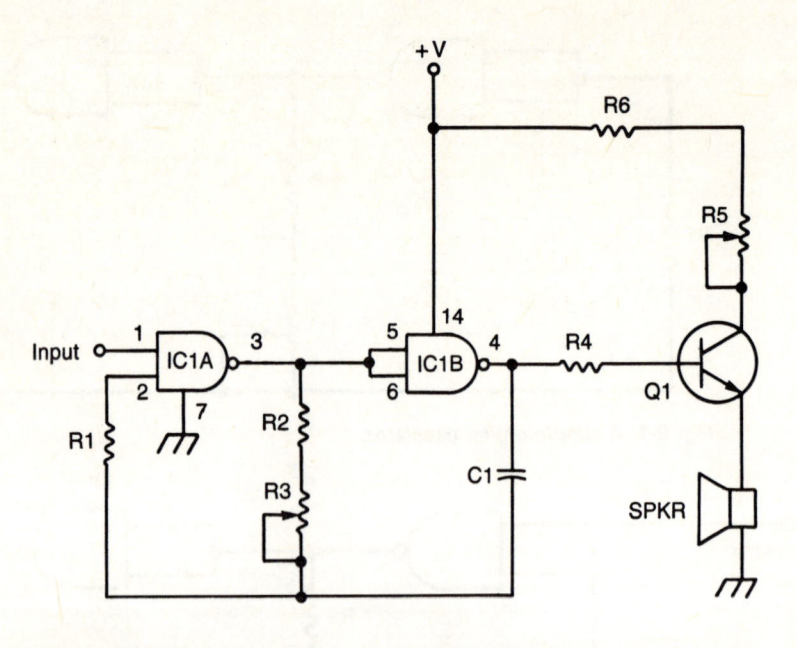

#20 PROJECT PARTS LIST	
COMPONENT	DESCRIPTION
IC1	CD4011 quad NAND gate
R1	1 MΩ resistor
R2	33kΩ resistor
R3	100 kΩ potentiometer
R4	12kΩ resistor
R5	500Ω potentiometer
R6	220Ω resistor
C1	0.05 µF capacitor (* see text)
Q1	NPN transistor (2N3904, 2N2222, or similar) (* see text)
SPKR	small loudspeaker (8Ω)

Project 21: Tunable Oscillator

Digital oscillators are usually pretty simple, but for a given set of components, they usually aren't tunable over a very wide range. Beyond certain limits they tend to be subject to stability problems.

We can construct a wide-range tunable digital oscillator by using a slightly esoteric chip—the CD4046 CMOS Phase Locked Loop IC. Phase Locked Loops (or PLLs) seem intimidating and mysterious to many hobbyists, and even a significant number of professional technicians. Actually PLLs aren't all that complicated. A PLL is basically just a VCO (Voltage-Controlled Oscillator) with feedback and an error detector. The output frequency is compared with a reference frequency. If the output frequency starts to drift, the error detector generates a correction voltage that is fed to the input of the VCO pulling its output back on frequency. This is not the place to go into detail on the operation of the Phase Locked Loop. Just don't let its *mystical* reputation scare you off.

Figure 6-4 shows a simple circuit using the CD4046 PLL as a tunable oscillator. Potentiometer R1 can adjust the output frequency over a fairly wide range. The circuit should be stable throughout its range. A typical parts list for this project is included with Fig. 6-4, but feel free to experiment with other component values.

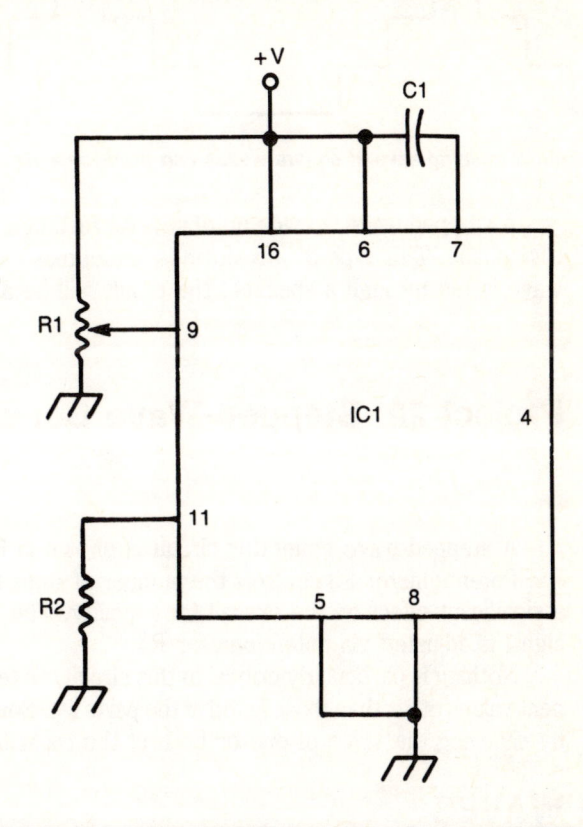

#21 PROJECT PARTS LIST

COMPONENT	DESCRIPTION
IC1	CD4046 CMOS PLL
C1	0.001 µF capacitor
R1	500kΩ potentiometer
R2	100kΩ resistor

Fig. 6-4. A PLL can be used as the heart of a tunable oscillator.

STEPPED-WAVES

Many circuits generate basic waveforms, including sine waves, triangle waves, sawtooth waves, and rectangle waves. Some applications call for more complex waveshapes. In music synthesis, the standard waveshapes can soon become rather boring.

One type of complex waveform that can easily be generated with digital circuitry is the stepped wave. There are many possible variations. A typical stepped wave is illustrated in Fig. 6-5.

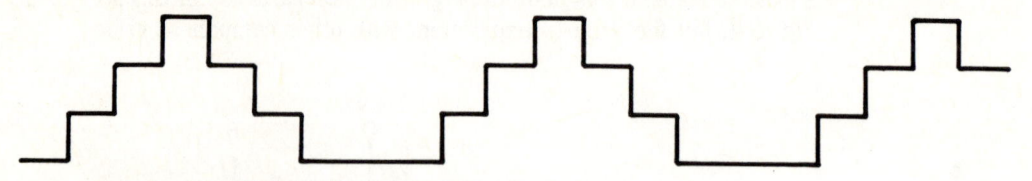

Fig. 6-5. Combining several square waves can produce a stepped wave.

A stepped wave is made up of several rectangle waves blended into a sort of staircase shape. This type of waveform is sometimes called a staircase wave. If a stepped wave is fed through a speaker, the result will be a very bright, brassy tone.

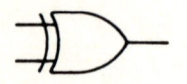

Project 22: Stepped-Wave Generator

A stepped-wave generator circuit is shown in Fig. 6-6 with its parts list.

Potentiometer R3 controls the number of steps in the output waveform, which has a significant effect on the overall tone quality. The frequency (or pitch) of the output signal is adjusted via potentiometer R4.

Nothing is particularly critical in this circuit, so feel free to experiment with component values other than those listed in the parts list. Some very odd effects can be achieved by changing the value of one or both of the capacitors.

TRIANGLE WAVES

Normally, digital circuits work with square and rectangle waves only. Since only two discrete states (HIGH and LOW) are recognized by logic gates, gradually changing waveforms are not necessary and could even cause problems within the digital circuitry. Even so, occasionally it might be desirable to synthesize an analog waveform via digital means.

Digital sine wave generation is covered elsewhere in this book. Being the purest analog waveform, the sine wave is the most difficult to synthesize digitally. In some applications we may not need a perfectly pure sine wave. Often we can "get away with" a triangle wave that contains all of the odd harmonics, although the harmonics are all quite low in level. In many noncritical applications, a triangle wave can be substituted directly for a sine wave. Low-pass filtering will result in something closer to an actual

Fig. 6-6. A stepped wave generator.

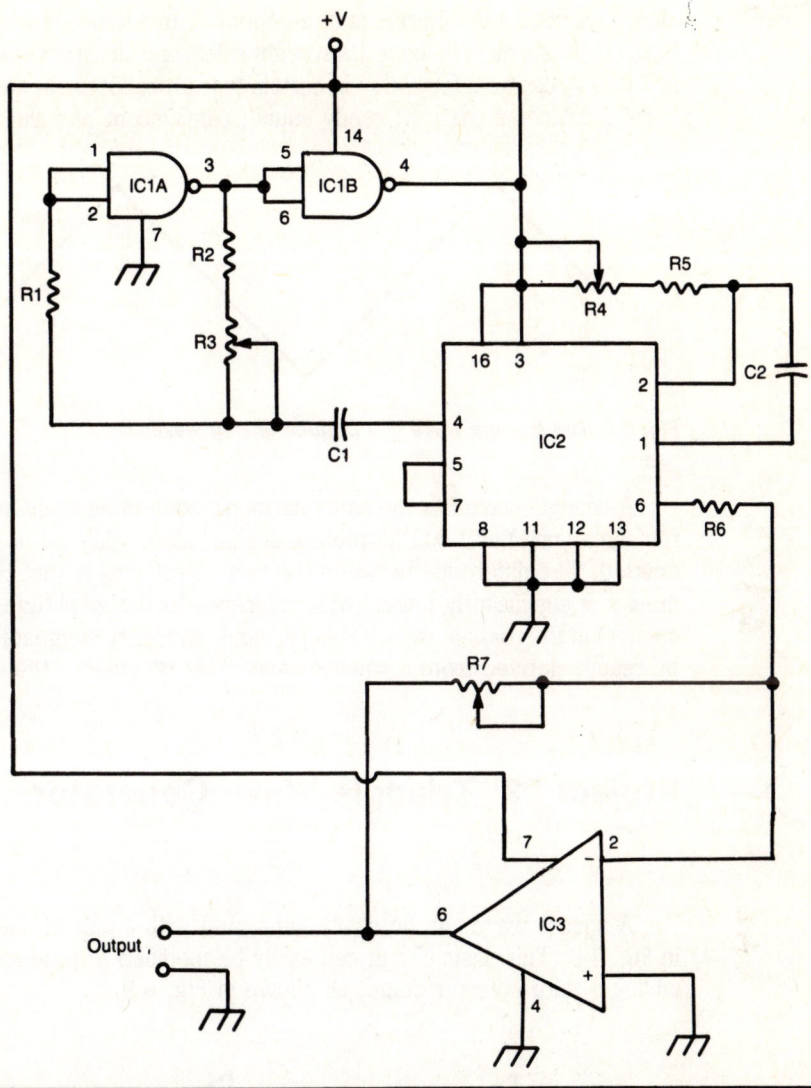

| #22 PROJECT PARTS LIST ||
COMPONENT	DESCRIPTION
IC1	CD4011 quad NAND gate
IC2	CD4528 dual one shot
IC3	op amp (741, or similar)
C1	0.01 µF capacitor *
C2	0.1 µF capacitor
R1	1 MΩ resistor
R2	47kΩ resistor
R3, R4, R7	100kΩ potentiometer
R5, R6	10kΩ resistor
(* see text)	

sine wave, but such filtering is often optional. In addition, the triangle wave has a number of applications of its own. Its straight sides, and sharp reversals of direction, as shown in Fig. 6-7, can be informative in certain test procedures. In electronic music, a triangle wave produces a pleasant reedy sound, reminiscent of a flute.

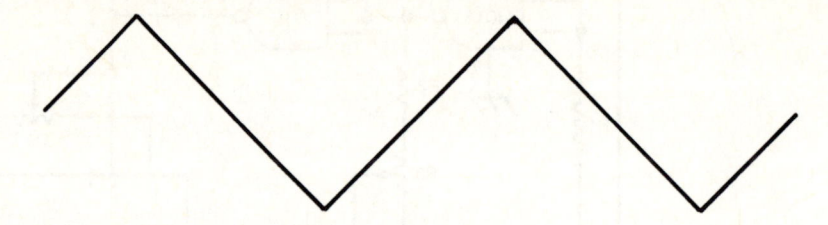

Fig. 6-7. The triangle wave is a popular analog waveshape.

A triangle wave has the same harmonic content as a square wave (50% duty cycle rectangle wave). All odd harmonics are included, while all of the even harmonics are omitted. The difference between the two waveforms is that the amplitude of the harmonics is significantly lower (with reference to the amplitude of the fundamental frequency) in the triangle wave. This harmonic similarity suggests that a triangle wave can be readily derived from a square wave. This is, indeed, the case.

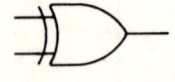

Project 23: Triangle-Wave Generator

A square wave can be easily generated with a pair of digital inverters, as shown in Fig. 6-8. This basic circuit can easily be modified to produce triangle waves just by adding a third inverter stage, as shown in Fig. 6-9.

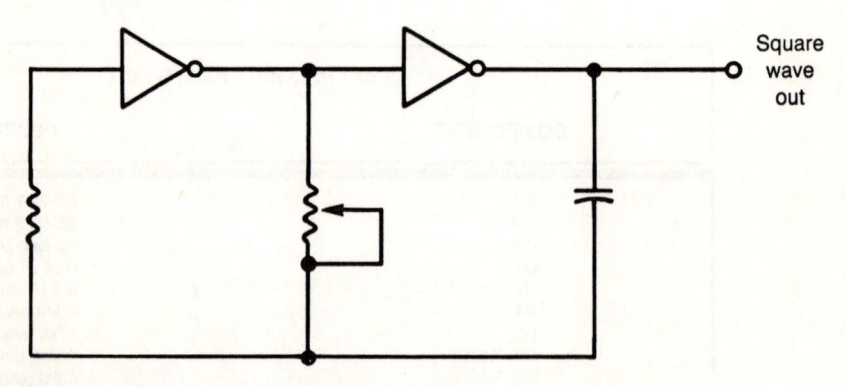

Fig. 6-8. Two inverters can generate a square wave.

Fig. 6-9. Adding a third inverter to the circuit of Fig. 6-8 creates a digital triangle wave generator.

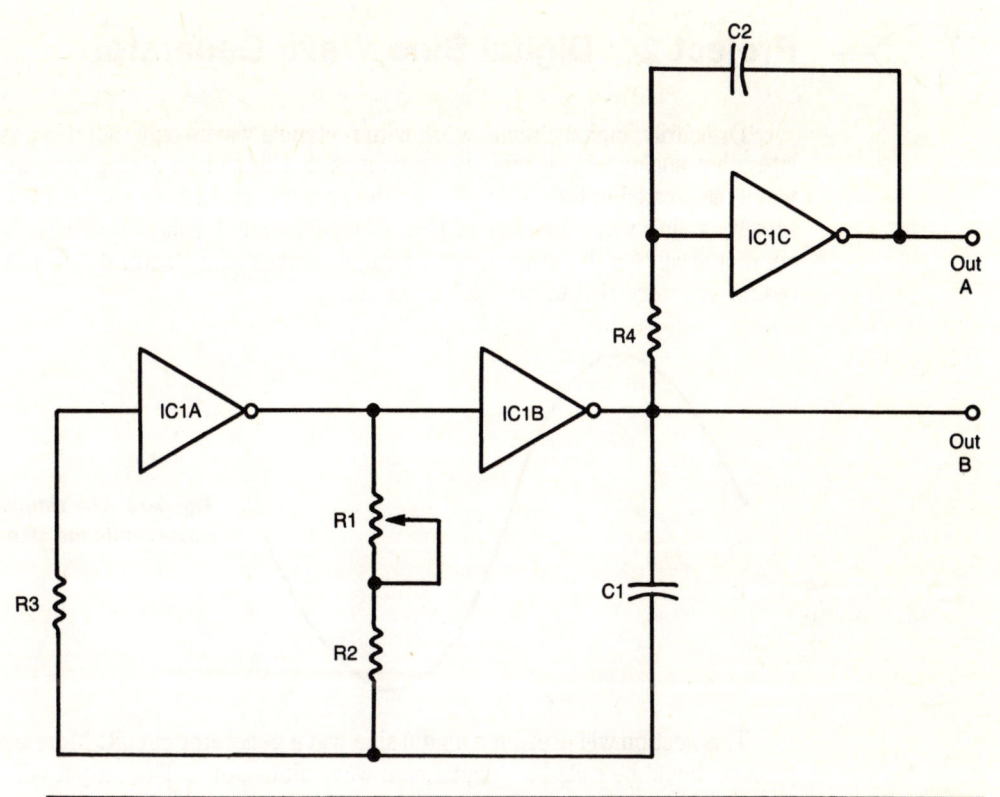

#23 PROJECT PARTS LIST	
COMPONENT	DESCRIPTION
IC1	CD4049 hex inverter
R1	100kΩ potentiometer
R2, R4	22kΩ resistor
R3	680kΩ resistor
C1, C2	0.047 µF capacitor

The "secret" of this circuit lies in the capacitive feedback path around the third inverter stage. Essentially, this capacitor functions as a simple low-pass filter, reducing the level of the harmonics.

In addition to the triangle wave output at point A, the original square wave output is simultaneously available at point B. Both signals, of course, will be at exactly the same frequency.

A typical parts list for this project is given with Fig. 6-9, but be sure to experiment with other component values. R1 and C1 control the signal frequency, while the value of C2 affects the exact shape of the triangle wave signal.

91

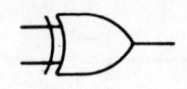

Project 24: Digital Sine Wave Generator

Ordinarily, digital circuits work with rectangle waves only, but it is possible to simulate other analog waveforms via digital means. The purest analog waveform is the sine wave, illustrated in Fig. 6-10. Notice the smooth, continuous curves of this waveform. Ideally, a sine wave consists of just its fundamental frequency. There is no harmonic content at all. On the other hand, rectangle waves (the "natural" signals of the digital realm) are very rich in strong harmonics.

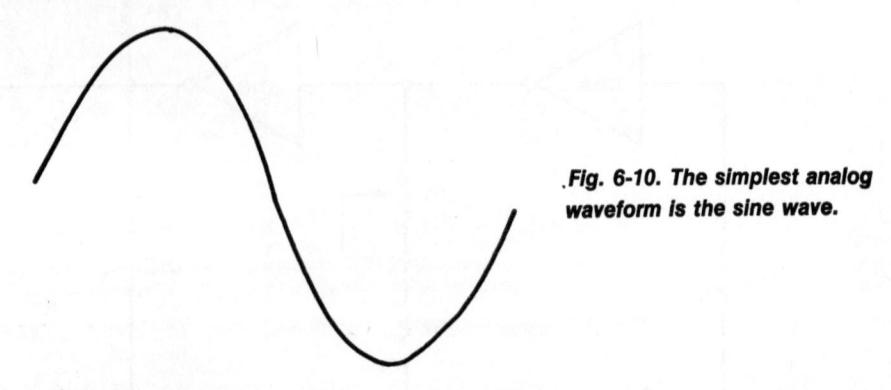

Fig. 6-10. The simplest analog waveform is the sine wave.

This section will present a digital sine wave generator circuit. More accurately, this project should be called a "digital sine wave simulator". The output will not be a true harmonic-free sine wave. There will inevitably be some harmonic content distorting the pure sine shape. The pseudo-sine wave generated by this circuit will contain only very weak, relatively high-order harmonics, and it will be close enough to a true sine wave for many practical purposes.

The sine wave generator circuit is built around a counter IC. Specifically, we will be working with the CD4018, which is called a programmable counter. The pin-out for this device is shown in Fig. 6-11.

The CD4018 can operate in either of two modes. In one mode, it acts like an ordinary counter, like the CD4017. In this mode the chip can divide (count) the input frequency (or pulse rate) by any whole number from 2 to 10. A feedback loop is required for this circuit to operate. The output is taken off pin #1 (DATA INPUT), which is in the feedback path. The output is a square wave (or a near square wave—odd counts can throw off the symmetry somewhat) that has a frequency determined by:

$$F_o = F_i/C$$

where F_o is the output frequency, F_i is the input frequency, and C is the count value.

For the CD4018's other major mode of operation, the JAM inputs (pins #2, 3, 7, 9, and 12) are used to program the starting count. The count can begin at any value between zero and the maximum count value. For example, you could set up a counting sequence like this:

```
3 4 5 6 7 8 3 4 5 6 7 8 3
4 5 6 7 8 3 4 5 . . .
```

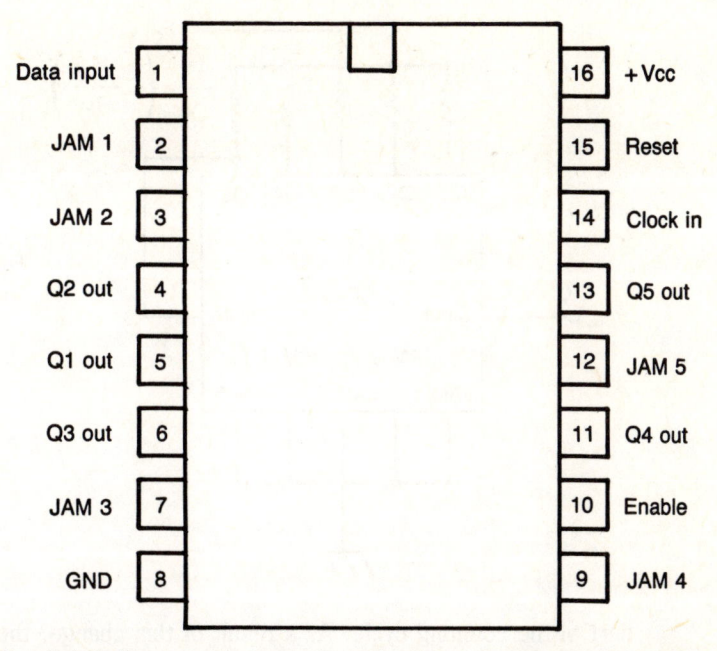

```
Data input   1          16   +Vcc
JAM 1        2          15   Reset
JAM 2        3          14   Clock in
Q2 out       4          13   Q5 out
Q1 out       5          12   JAM 5
Q3 out       6          11   Q4 out
JAM 3        7          10   Enable
GND          8           9   JAM 4
```

Fig. 6-11. The digital sine wave generator is built around the CD4018 counter IC.

The jam inputs are loaded in parallel fashion, much like loading a parallel input shift register. As with most CMOS devices, all unused input pins on the CD4018 should be grounded.

The digital sine wave generator is based on the fact that the CD4018 provides several phase-shifted outputs. Each one is delayed from its predecessor by exactly one input (clock) pulse. If we sum the outputs together with the correct relative weighting, the result will be a staircase wave, like the one illustrated in Fig. 6-12. External low-pass filtering can smooth out the steps to create a pseudo-analog waveform. The basic circuit is shown in Fig. 6-13.

The actual output waveshape is defined by the relative weighting of each individual output. This weighting, in turn, is controlled by the resistor values. If the resistors have equal values, the filtered output will more or less resemble a triangle wave.

In a sine wave, the peaks are fairly flattened out. A triangle wave, and (even more so) the staircase waveform produced by the counter have rather sharp peaks. Figure 6-14 shows a modification of the basic circuit to produce a waveform with flatter peaks. This is done by eliminating output Q5 from the system output, even though it is still

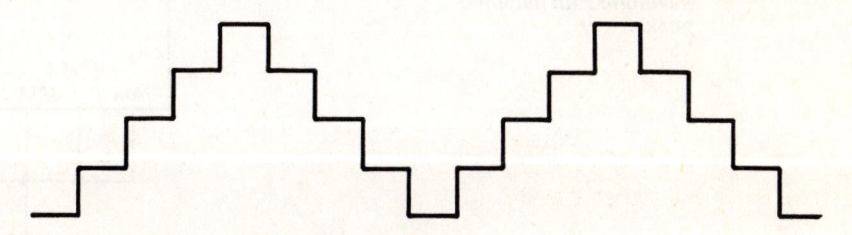

Fig. 6-12. A digital approximation of a sine wave.

$$F_{OUT} = \frac{F_{IN}}{10}$$

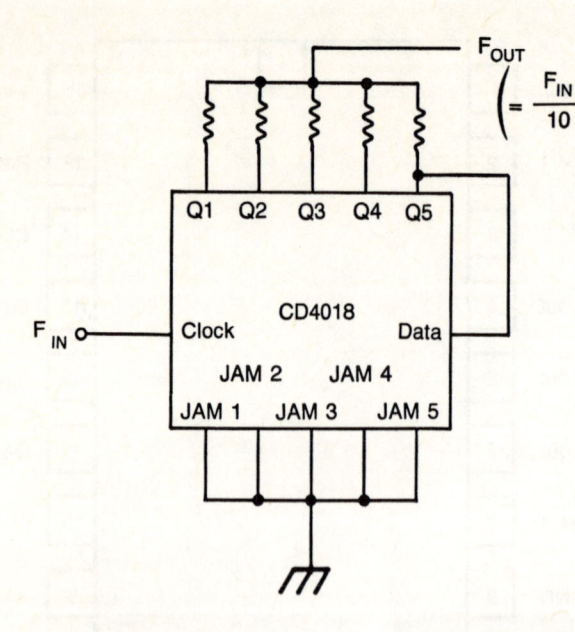

Fig. 6-13. The basic digital sine wave generator circuit.

part of the counting cycle. As a result of this change, the Q4 peak is effectively held for a longer period of time. Filtering this signal gives us a reasonably smooth pseudo-sine wave.

As always (or at least, so it seems) this improvement comes at a price. Since one less counting output is used in creating the output waveform, the resolution is reduced. Instead of five steps to the peak, there are only four. This could be a problem, depending on the specific application at hand.

A practical digital sine wave generator circuit using the principles discussed here is shown in Fig. 6-15, along with its parts list for this project.

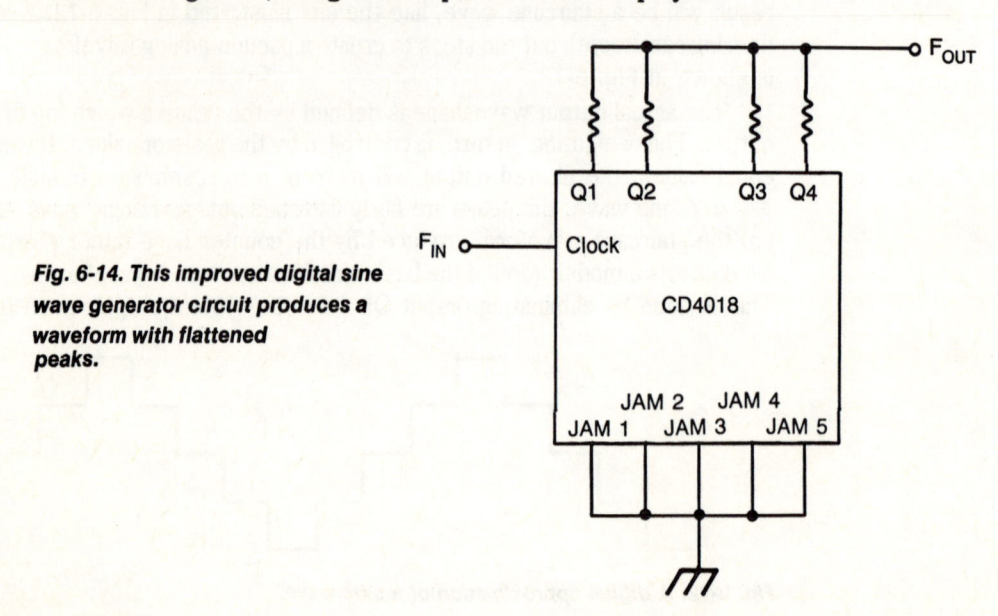

Fig. 6-14. This improved digital sine wave generator circuit produces a waveform with flattened peaks.

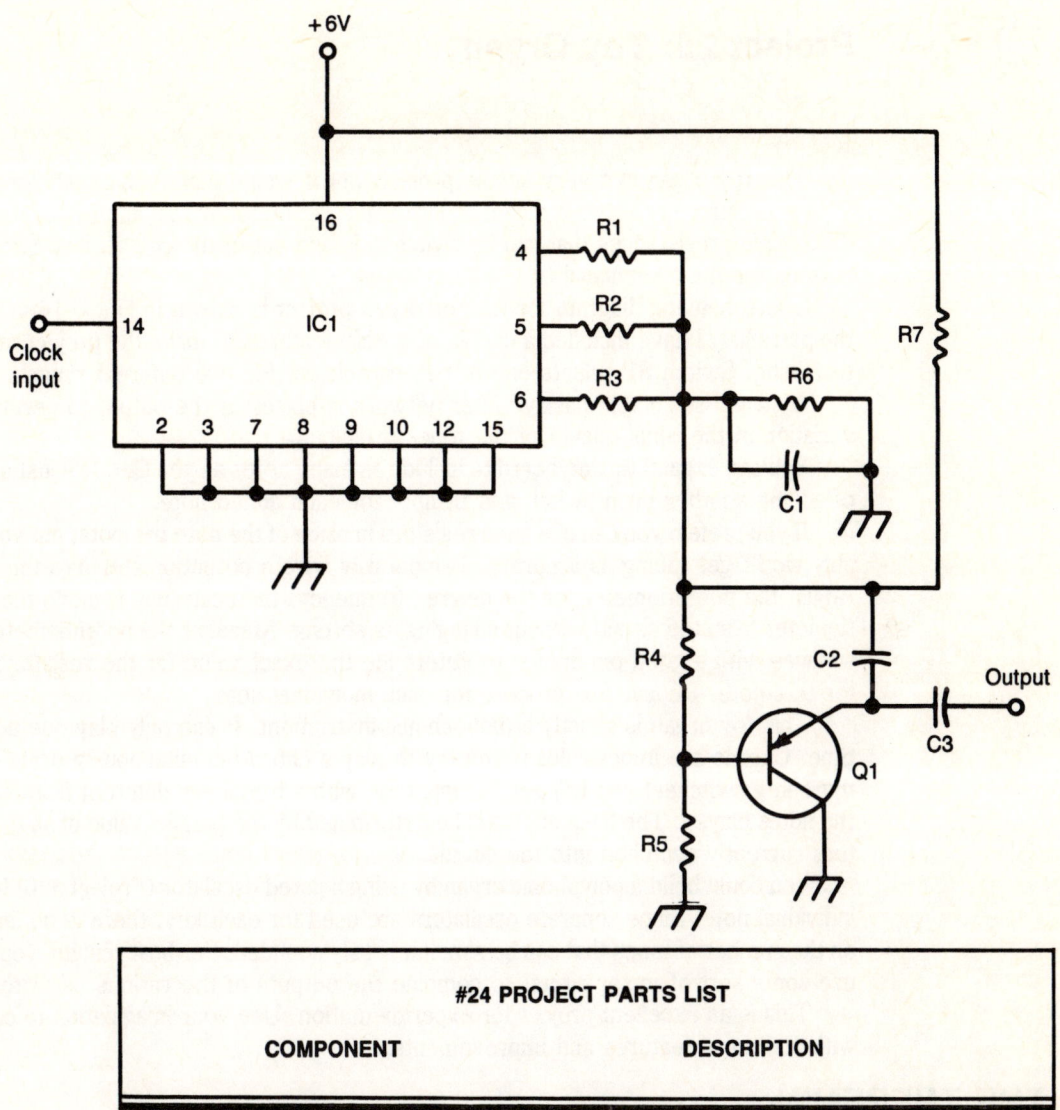

#24 PROJECT PARTS LIST

COMPONENT	DESCRIPTION
IC1	CD4018
Q1	PNP transistor (2N3906, or similar)
C1	0.0056 µF capacitor
C2	0.0015 µF capacitor
C3	0.5 µF capacitor
R1, R7	2.2kΩ resistor
R2	30kΩ resistor
R3	22kΩ resistor
R4, R5	33kΩ resistor
R6	4.7kΩ resistor

Fig. 6-15. A practical digital sine wave generator circuit.

Project 25: Toy Organ

This toy organ is a very simple project, but it would make a nice gift for a child. This is nothing more than a digital square wave generator with a number of frequency determining resistances that can be switched in and out of the circuit. Use SPST push buttons for the "keyboard".

The schematic diagram for the toy organ project is shown in Fig. 6-16 as well as the parts list. I have included a couple of special features to make the project more entertaining. Switch S1 selects one of two capacitors, for two different ranges.

A switch-selectable passive filter network is placed at the output to permit some variation in the tonal quality of the tones produced.

You can expand the keyboard to include as many notes as you like. It's just a matter of adding another push switch and trimpot for each added note.

If you prefer, you can use fixed resistors in place of the note trimpots, but you probably won't get tuning as accurate. Temporarily wire a potentiometer into the circuit. Adjust the potentiometer for the desired frequency, then carefully remove the potentiometer from the circuit without changing its setting. Measure the potentiometer's resistance with a good ohmmeter to determine the exact value for the resistor needed for that note. Repeat this process for each individual note.

This toy organ is strictly a monophonic instrument. It can only play one note at a time. Chords are impossible. If you try to play a chord (simultaneously press two or more key switches) you will get a single tone with a frequency different from either of the notes played. The frequency will be determined by the parallel value of all the resistors currently switched into the circuit.

You could build a polyphonic organ by using a gated oscillator (Project #20) for each individual note. Since separate oscillators are used for each key, there is no limitation on the number of tones that can be simultaneously sounded. For best results, you should use some sort of mixer circuit to combine the outputs of the various oscillators.

This is an excellent project for experimentation. Use your imagination to come up with additional features and improvements.

THE THEREMIN

Before the modern music synthesizer was developed a surprising number of electronic musical instruments were invented. One of the most unusual, and still intriguing is the theremin. This odd instrument was the brainchild of a Russian by the name of Leon Theremin.

The strange thing about the theremin is that the performer plays it without even touching it. The performer waved his hands over one or two antennas to produce music. The capacitance of the player's hands was sufficient to change the resonant frequency of delicate tuned circuits within the instrument.

Not surprisingly, the theremin was difficult to tune, and even more difficult to play. Ambient conditions could seriously affect the tuning of the device. Still the idea of "no-touch" playing is fascinating, and to a certain type of mind, almost irresistible.

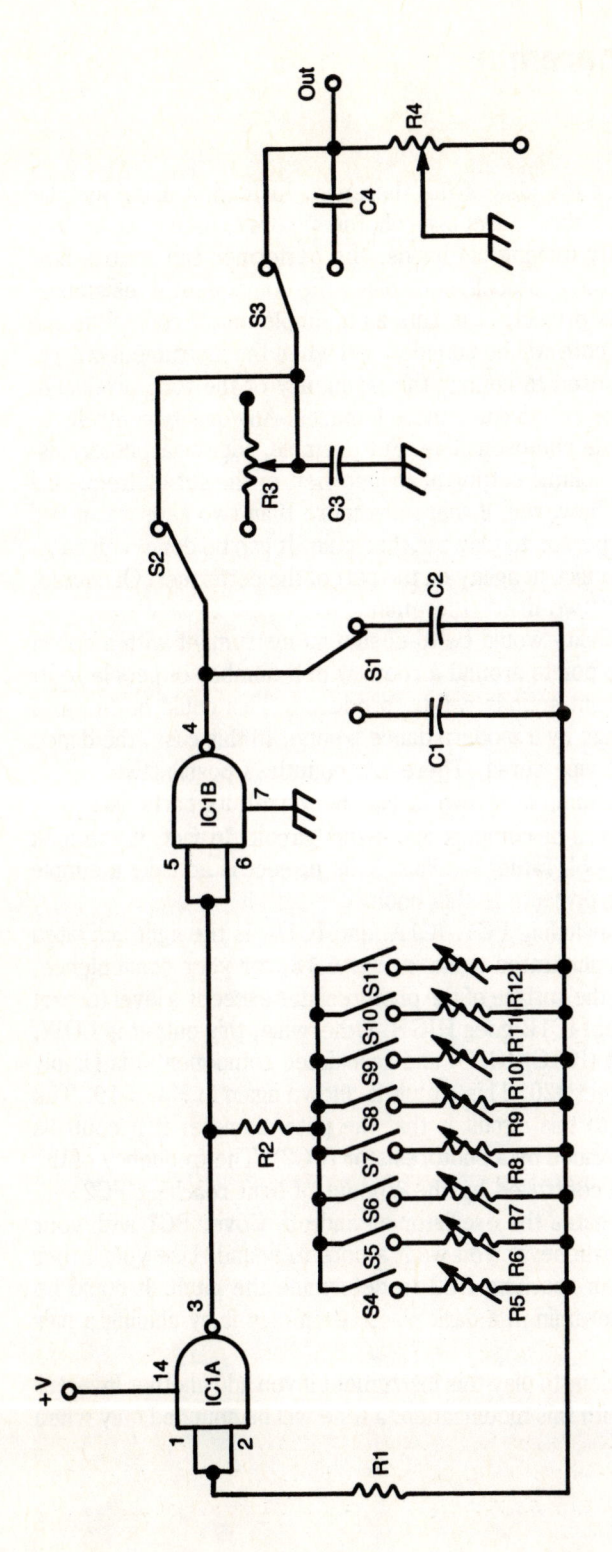

#25 PROJECT PARTS LIST

COMPONENT	DESCRIPTION
IC1	CD4011 quad NAND gate
C1	0.01 µF capacitor
C2	0.047 µF capacitor
C3, C4	0.1 µF capacitor
R1	1 MΩ resistor
R2	10kΩ resistor
R3, R4	500kΩ potentiometer
R5 - R12	250kΩ trimpot (*)
S1, S2, S3	SPDT switch
S4 - S11	Normally Open SPST push switches (*)

(*) add more for more notes—see text

Fig. 6-16. The schematic for the toy organ project.

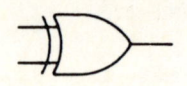

Project 26: Photo-Theremin

In this project I have updated the idea of the theremin. Instead of using metallic antennas to sense hand capacitance, this project uses photoresistors to detect the amount of light falling on their surface. By moving his hands, the performer can control how much light and shadow will reach the photocell, controlling the component's resistance.

Two sensors are used in this project. One acts as a simple on/off control so the performer can determine when a note will be sounded and when the instrument will remain silent. The other sensor is used to control the frequency of the tone produced. You might want to expand this project to include more features. Any quality controllable by a variable resistance can be made photosensitive. For example, additional photoresistors could be added to control the volume setting of an amplifier, or the cut-off frequency of a filter. One word of warning, however, if there are more than two sensors, it will be extremely difficult for any one person to play the theremin. It can be done with careful placement of the sensors, and sufficient agility on the part of the performer. Of course, it might be interesting to play a "two-man" theremin.

An even more extreme possibility would be to design an instrument with a dozen or so sensors, placed at strategic points around a room with a number of people in it. As the people move about, different sounds will be produced. This could be fun at a party. It could also be used on stage by a modern dance troupe. In this case, the dance determines the music, instead of vice versa. There are countless possibilities.

The circuit for the photo-theremin is shown in Fig. 6-17 with its parts list.

There is nothing terribly critical or complex about this circuit. In fact, if you look closely at a schematic, it should look rather familiar. This project is actually a simple combination of two of the earlier projects in this book.

The first part of the circuit (including PC1, IC1A, and IC1B) is the light activated gate of Project #8. Project #8 is illustrated again in Fig. 6-18, for your convenience. When the amount of light reaching the surface of the photoresistor exceeds a level (preset by potentiometer R1), the output of IC1B goes HIGH. Otherwise, this output is LOW.

The second half of the circuit (IC1C, IC1D, and associated components) is simply the gated oscillator circuit of Project #20. This circuit is shown again in Fig. 6-19. The only change that has been made to this circuit is that the potentiometer that controls the output frequency has been replaced by a photoresistor (PC2). The frequency of the tone produced by the speaker is controlled by the amount of light reaching PC2.

The light or shadow on PC1 gates the oscillator on and off. Cover PC1 with your hand, and move your hand away whenever you want a note to sound. Use your other hand to control the amount of light reaching PC2 to determine the pitch. It could be interesting to place your phototheremin in a dark place, then play it by shining a pair of flashlights on the sensors.

You might find it more convenient to play this instrument if you add another inverter stage between IC1B and IC1C. With this modification, a tone will be sounded only when PC1 is covered (in shadow).

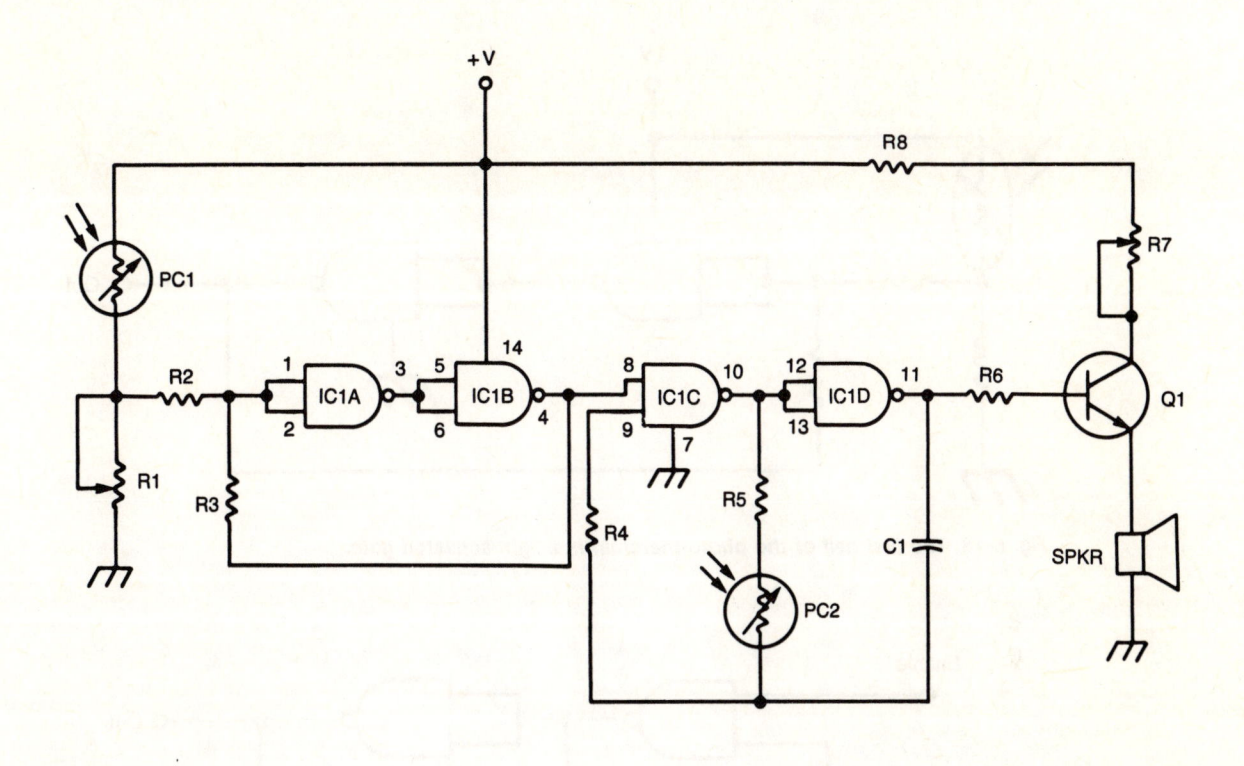

Fig. 6-17. The circuit for the photo-theremin project.

#26 PROJECT PARTS LIST

COMPONENT	DESCRIPTION
IC1	CD4011 quad NAND gate
Q1	NPN transistor (2N3904, 2N2222, or similar)
PC1, PC2	photoresistor
R1	50kΩ potentiometer
R2, R6	10kΩ resistor
R3	100kΩ resistor
R4	1 MΩ resistor
R5	1kΩ resistor
R7	500Ω potentiometer
R8	220Ω resistor
C1	0.01 µF capacitor
SPKR	small loudspeaker (8Ω)

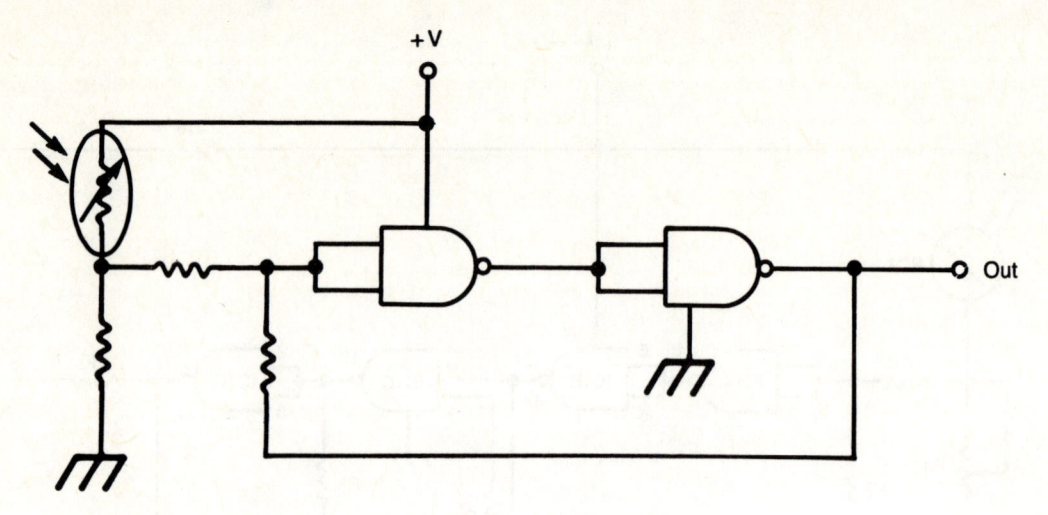

Fig. 6-18. The first half of the photo-theremin is a light-activated gate.

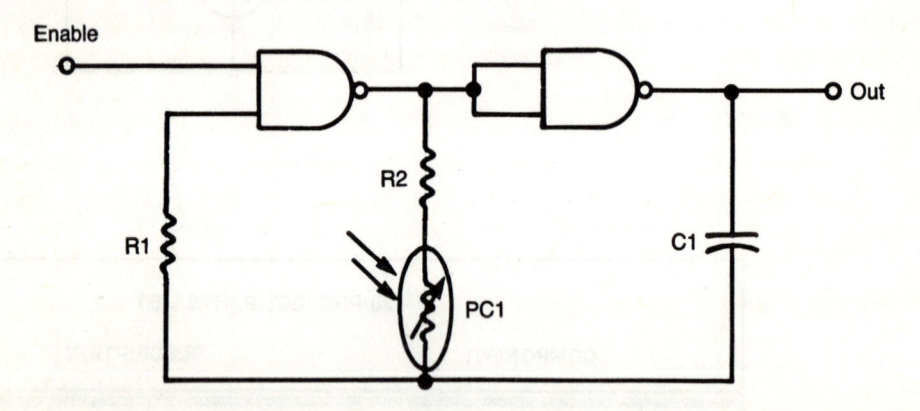

Fig. 6-19. The tone from the photo-theremin is produced by a variation of the gated oscillator.

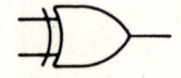

Project 27: Four Tone Sequencer

I have always had a particular fondness for musical projects, especially musical circuits that "play themselves." Automated music makers are fascinating, and a lot of fun.

Digital circuitry is ideal for automated music makers. Digital counters make dandy sequencers to activate a series of tones.

There are a number of different ways to sound a specific tone when a specific point in the sequence is reached. Perhaps the most direct method is to use a gated oscillator, like the one shown in Fig. 6-20. This circuit should look familiar. We have used it in a number of the projects in this book. The oscillator can generate a signal only when

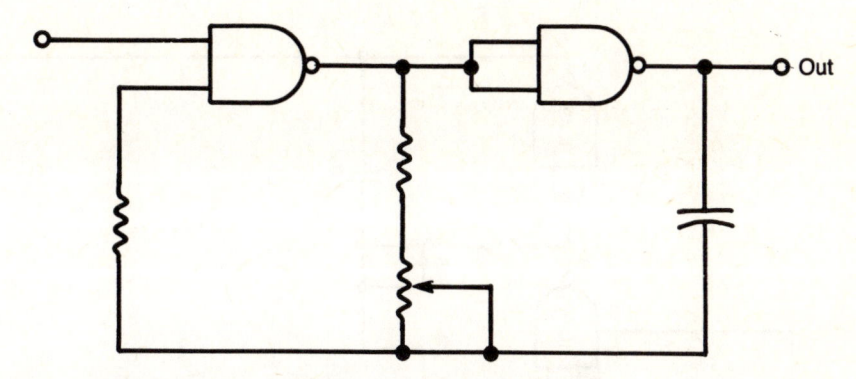

Fig. 6-20. A sequencer can selectively activate gated oscillators.

there is a HIGH signal on the ENABLE input. If the ENABLE input is held LOW, the oscillator will be cut off.

One gated oscillator is used for each step in the sequence. This would become unwieldy and rather expensive for sequencers with many steps, but for a small sequencer this is a very practical method.

In this project we will combine four of these gated oscillators with the four-step sequencer circuit of Project #47. The complete schematic for this project is illustrated in Fig. 6-21. The parts list is included.

While the schematic looks a little complicated this is really a fairly simple project. Much of the apparent complexity comes from the fact that the gated oscillator circuit is repeated four times.

Potentiometer R2 controls the speed of the sequence. The other four potentiometers in the circuit (R5, R8, R10, and R15) are used to set the frequency of each of the gated oscillators. Any four tones (notes) can be set up. Sometimes you can achieve some very interesting results by retuning one or more of the oscillators while the sequence is playing.

Once the five potentiometers are set, and power is applied to the circuit, a four note pattern will be repeated over and over.

This circuit is designed so that only one of the gated oscillators will ever be activated at any given instant, there is no need for an output mixer stage. The four oscillator outputs can simply be tied together, as shown in the schematic.

If things start to get a little boring, try substituting other component values for any of the passive components. Nothing is particularly critical in this circuit. I will offer one suggestion, however, don't bother experimenting with the value of capacitor C2. This component just helps to stabilize the 7555 timer. In many cases it isn't really needed, but it's cheap insurance against possible stability problems. In any event, changing the value of C2 will not affect circuit operation in any noticeable way. All of the other passive components in this circuit will have some practical effect on the output tone sequence.

The components associated with IC1 (R1 through R3, and C1) control the speed of the sequence. The other passive components control the frequency of one of the oscillators. Since the circuit already has potentiometers to adjust the resistances, you'll get the most interesting results if you experiment with the capacitor values (C1, C3, C4, C5, and C6).

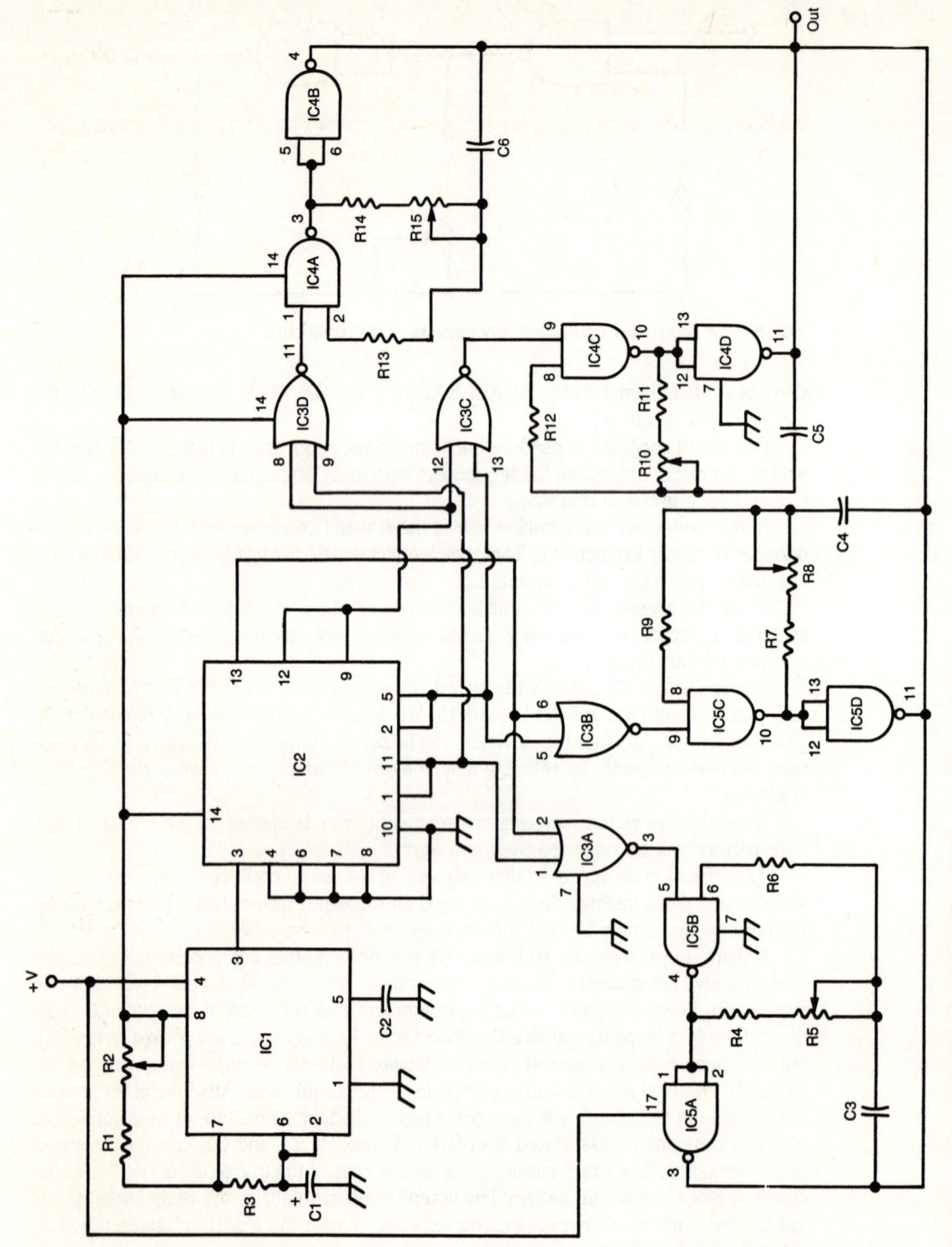

Fig. 6-21. The complete schematic for the four tone sequencer project.

```
                    #27 PROJECT PARTS LIST

         COMPONENT                        DESCRIPTION

         IC1                              7555 timer
         IC2                              CD4013 dual flip-flop
         IC3                              CD4001 quad NOR gate
         IC4, IC5                         CD4011 quad NAND gate
         C1                               25 µF electrolytic capacitor
         C2 - C6                          0.01 µF capacitor
         R1                               1.2kΩ resistor
         R2, R5, R8, R10, R15             500kΩ potentiometer
         R3                               4.7kΩ resistor
         R4, R7, R11, R14                 22kΩ resistor
         R6, R9, R12, R13                 1 MΩ resistor
```

Fig. 6-21. (continued)

Some fascinating effects can be achieved if you operate two of these circuits simultaneously at just slightly different speeds.

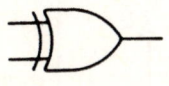

Project 28: Ten Step Tone Sequencer

In Project #27 we built a four step tone sequencer. Each of the four tones was generated by its own individual gated oscillator. The control circuitry turned the oscillators on and off in sequence. For just four separate tones, this is a reasonable approach. But for longer tone sequences using separate gated oscillators becomes increasingly awkward. An alternative approach is to use a single tone generator. In this case, the control circuitry selects one of several available frequency determining components—usually resistors or capacitors. Resistors are generally the better choice, because inexpensive variable resistors (trimpots) are available. This makes it easy to fine tune the individual tones in the sequence.

The CD4066 quad bilateral switch is a good choice for controlling the oscillator. Any individual component can be switched into the circuit under digital control. A single switch from the CD4066 is illustrated in Fig. 6-22. Notice that there are three connections.

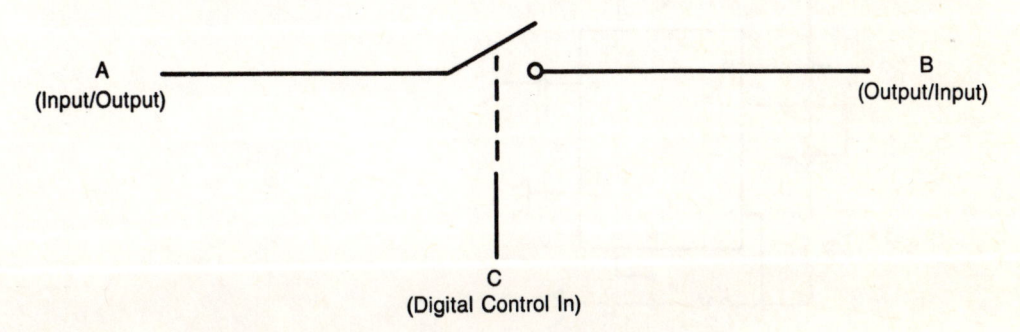

Fig. 6-22. The CD4066 contains four digitally controllable switches.

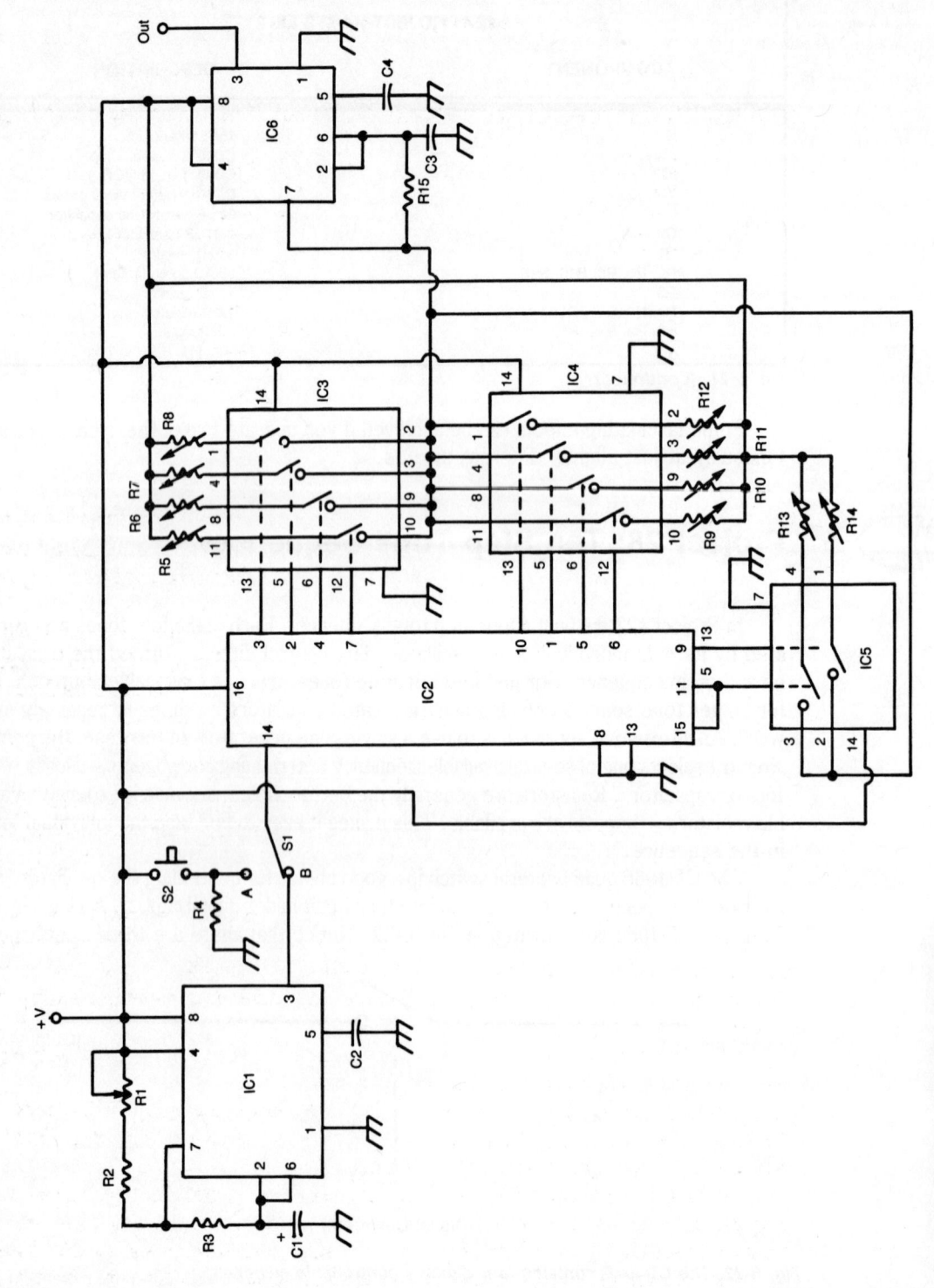

Fig. 6-23. The ten tone sequencer circuit.

```
                    #28 PROJECT PARTS LIST

       COMPONENT                        DESCRIPTION
==============================================================
       IC1, IC6                7555 timer
       IC2                     CD4017 decade counter
       IC3, IC4, IC5           CD4066 quad bilateral switch
       C1                      47 µF electrolytic capacitor (*)
       C2, C4                  0.01 µF capacitor
       C3                      0.1 µF capacitor (*)
       R1                      100kΩ potentiometer
       R2, R3                  2.2kΩ resistor
       R4                      1 MΩ resistor
       R5 - R14                100kΩ trimpot (*)
       R15                     10kΩ resistor
       S1                      SPDT switch
       S2                      N.O. SPST push-switch

       (*) see text
```

Connection points A and B are treated exactly the same as for a regular mechanical switch. Either end can be used as input or output. The switch is bidirectional. (This is why the word *bilateral* is included in this IC's name.)

The third connection point (C) is a digital control input. If this input is held LOW, the switch is open. The switch is closed by raising the control input to the HIGH state.

A CD4066 chip contains four of these switches. They are independent of each other. The only thing they share is the power supply connections to the IC.

For more information on the CD4066, refer back to Project #12—The Programmable Resistance/Capacitance Network.

Most digital sequencers use a counter to step through the series of outputs. If we use a decimal counter like the CD4017 we can build a ten step sequencer. By shorting pin #15 (RESET) to pin #11 (output 9), the sequence will endlessly repeat itself as long as power is applied to the circuit.

A complete schematic of a ten step tone sequencer project is shown in Fig. 6-23. This circuit looks a little complex. Six ICs are used. But the way each IC is used is simple, mirroring other, simpler projects in this book. If you consider each IC (and its associated components) individually, you should have no trouble at all understanding the way this circuit functions.

IC1 is the system clock. It is nothing but a simple astable multivibrator (rectangle wave generator) built around a 7555 (or 555) timer IC.

Normally switch S1 will be in position B, connecting the output of IC1 (clock) to the input of IC2 (counter). Each clock pulse advances the counter one step. For the time being we will ignore position A of switch S1.

Adjusting potentiometer R1 controls the speed of the sequence. You might also want to consider experimenting with other values of capacitor C1; this has an effect on the stepping speed. If you use a high frequency clock rate (near or within the audible range), the various tones will blend together into a single complex tone, converting the project into a complex tone generator. The smaller the value of C1, the faster the clock will cycle the counter through its outputs. Perhaps you could include a range selector switch with several different capacitor values in place of the single C1 shown in the diagram.

IC2 is a CD4017 decimal counter, it cycles through its ten outputs, advancing one step each time its clock input (pin #14) is triggered. When it reaches a count of 9, it will reset itself back to 0, and start over. This is accomplished by connecting the RESET input (pin #15) to the 9 OUTPUT (pin #11). For a shorter sequence, short pin #15 to the highest desired output pin.

Pin #13 (HALT) is shown permanently shorted to ground in the diagram. This puts a constant logic 0 on this pin. You could adapt the circuit by using a switch to select between shorting this pin to ground (logic 0) and to +VDD (logic 1). A logic 1 at pin #13 will cause the counter to stop the sequence, holding the current output count value. To permit the counter to cycle through the sequence, this pin must be LOW.

Of IC2's ten outputs, only the one representing the current count value will be HIGH, the other nine outputs will be LOW.

The ten counter outputs are used to control the switches in two and a half CD4066 quad packages (IC3, IC4, and IC5). I'll leave it to you to come up with a use for the remaining two switches in IC5. Of course, you don't have to use them. Remember that *all* unused CMOS inputs should be shorted to ground or +VDD.

One end of the switches is shorted to all the others. An individual trimpot (adjustable resistance) is connected to the other end of each switch (R5 through R14). A logic 1 on the appropriate control input switches that particular resistor into the circuit. Because the counter activates only one output at a time only one switch will ever be closed, so there will be no problems with parallel values of the resistances.

If cost or space is a major concern, you could use fixed resistors in place of the trimpots, but this will eliminate the advantage of tunability.

The selected resistance will be used as part of the circuitry surrounding IC6, which is simply another 7555 astable multivibrator. The selected resistance will determine the frequency of the signal generated by this oscillator. Once again, you could add a range switch to select different values for capacitor C3.

The output of IC3 can be fed to an amplifier or a small loudspeaker. As long as power is applied to the circuit you will hear a repeating sequence of ten tones.

Now turn your attention back to switch S1. In position B, this switch allows the clock (IC1) to drive the counter (IC2). Moving switch S1 to position A cuts the clock out of the circuit. Switch S2 is a Normally Open SPST pushbutton. When it is open, pin #14 of IC2 is grounded through resistor R4. This puts a logic 0 signal on the counter's clock input. Momentarily depressing S2 applies the +VDD voltage to pin #14 of IC2, acting as a logic 1 signal, triggering the counter to advance one step. This "manual clock" can be used to make tuning the individual tones easier. You can step through the sequence one by one, taking as much time as you need to adjust each trimpot.

A word of caution is due here. A simple push-switch is prone to bouncing. What you thought was just one clean switch closing and opening, might look to the counter like a string of several discrete clock pulses, advancing the sequence several steps instead of just one. This is not disastrous in this type of application, but it can be annoying and frustrating.

Switch bouncing will most likely be a problem with inexpensive push-switches.

If the bouncing is too much of a problem there are several ways you can get around it. One way is to place LEDs at the counter (IC2) outputs, to indicate which output is HIGH. Be sure to use a current-dropping resistor, as shown in Fig. 6-24. Now, even if the switch bounces, and the counter skips ahead several steps, you will know which trimpot to tune. These LEDs could be connected temporarily, or, if you happen to like flashing lights, you could leave them as a permanent part of the project.

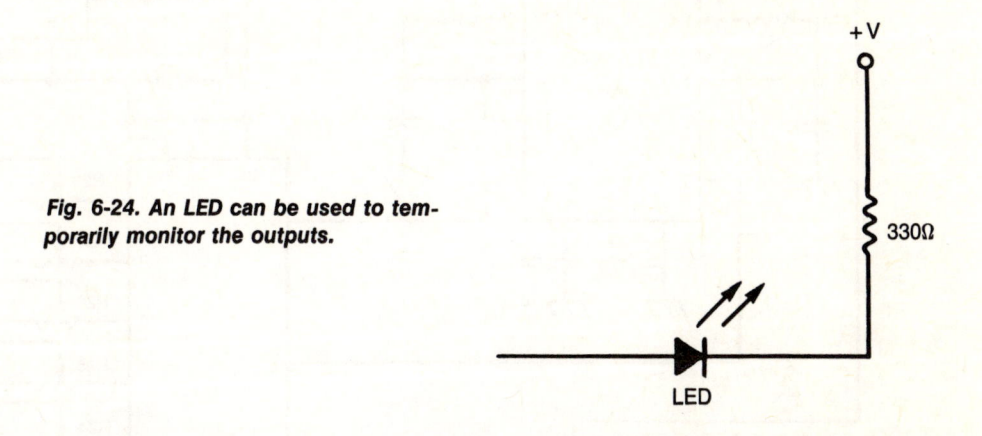

Fig. 6-24. An LED can be used to temporarily monitor the outputs.

A more elegant solution to the problem, of course, would be to use a switch debouncer circuit with S2. Such a circuit is featured in Project #43 of this book. A suitable parts list for this project appears in Fig. 7-23.

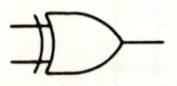

Project 29: Random Tone Generator

Sequencer project circuits play a preprogrammed series of tones over and over. While intriguing, the inherent redundancy can rapidly become stupefyingly boring.

If you are interested in this type of project, you'll surely be excited about a circuit that essentially "makes its own tunes" as it plays. Sort of an electronic composer.

This project can be fun, and educational, but don't expect too much. Each note is selected randomly, and not according to any rule of musical composition. For the most part the results will sound as random as they are, but that can be entertaining in itself, and every once in awhile you're sure to hear a snatch of melody.

The schematic for the random tone generator project is shown in Fig. 6-25 with the parts list.

The operator preselects sixteen tones by adjusting potentiometer R13 through R28. The adjustment is best done by ear. Set all of the other potentiometers to one extreme end of their scale so the note you are working on will stand out. It takes quite a bit of patience to make any precise adjustments. If you don't want to bother, just adjust each of the potentiometers to a different setting.

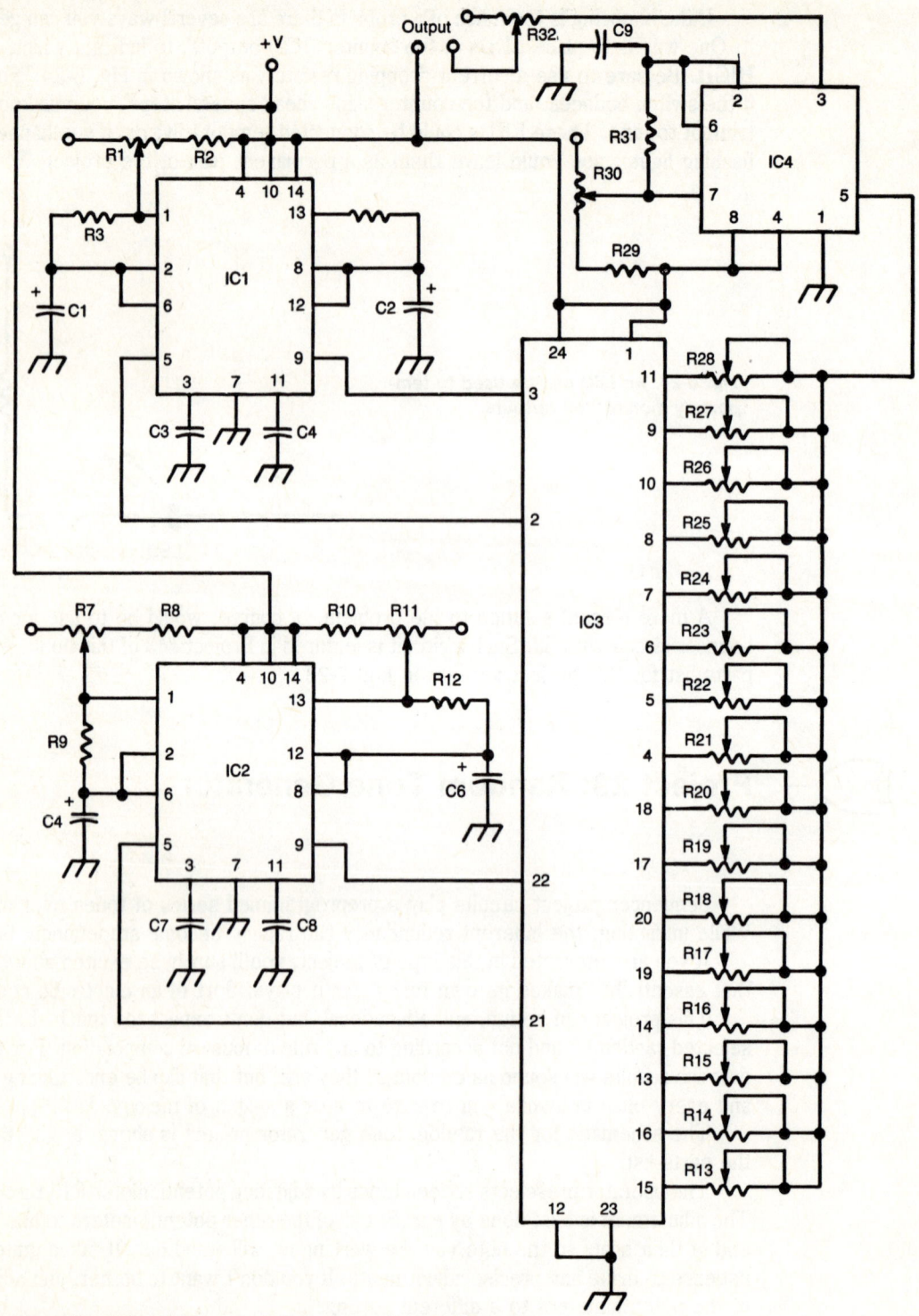

Fig. 6-25. This circuit generates a series of random tones.

#29 PROJECT PARTS LIST

COMPONENT	DESCRIPTION
IC1, IC2	556 dual timer (see text)
IC3	CD4514 4-to-16 Demultiplexer
IC4	7555 CMOS timer
C1	33 µF 35V electrolytic capacitor
C2, C5	47 µF 35V electrolytic capacitor
C3, C4, C7, C8	0.01 µF disc capacitor
C6	100 µF 35 volt electrolytic capacitor
C9	0.1 µF disc capacitor
R1, R5, R7, R11, R13 - R28	10kΩ potentiometer (or trimpot)
R2, R4, R8, R10	22kΩ resistor
R3, R6, R9, R12, R29	10kΩ resistor
R30	100kΩ potentiometer
R31	27kΩ resistor
R32	5kΩ potentiometer

If you'd like to keep the cost and/or the size of the project down, you could replace potentiometers R13 through R28 with fixed resistors. No two resistors should have the same value.

Essentially, these potentiometers determine the voltage that will be fed to a VCO built around a 7555 timer (IC4). The various potentiometers (and hence, their tone determining voltages) will be selected in random order as long as power is applied to the circuit.

The heart of this circuit is IC3. This chip is a CD4514 four-to sixteen demultiplexer. A four-bit binary number that is fed to the input of this chip causes it to activate one (and only one) of its sixteen outputs, putting a voltage through the appropriate potentiometer.

The four-bit binary control number is generated by four independent 555 timer oscillators. The CMOS 7555 timer could be used here, but the standard 555 units will work just as well. Parts count for the project can be reduced by using two 556 dual timer ICs in place of four independent timer chips. This is the method shown in the schematic.

The output of each of these oscillators serves as one of the bits in the control number. The four oscillators should be set up to run at independent rates (preferably very low rates) by selecting the following component values from Table 6-1:

Because each control oscillator has a different time base a very irregular pattern of binary numbers will be presented to the inputs of the demultiplexer. The following

OSCILLATOR NUMBER	COMPONENT VALUE EQUATION
1	$1.1 R_1 C_1$
2	$1.1 R_5 C_2$
3	$1.1 R_7 C_5$
4	$1.1 R_{11} C_6$

Table 6-1.

potentiometers also control the oscillator frequencies (actually, in the above equations the potentiometer value should be added to the resistance):

R1; R5; R7; R11

By adjusting these potentiometers you can roughly control how long each note will be held.

Finally, potentiometer R30 serves as a master pitch control to set the range of the device, and potentiometer R32 is an output volume control.

DIGITAL FILTERS

Digital circuitry can mimic many analog functions. One common type of analog circuit is the filter. A filter is a frequency selective circuit that allows certain frequencies to pass through to the output, but partially or completely blocks other frequencies.

Digital filters are now possible. They are most often used in communications systems, computer analysis of sounds, and music synthesizers.

To understand digital filters, we first need to examine basic analog filters. The most basic analog filter circuit is illustrated in Fig. 6-26. It is made up of just two passive components—a resistor and a capacitor. This is a low-pass filter. It offers greater attenuation (blocking) to higher frequencies. Essentially what happens is that high frequency signals are shunted through the capacitor to ground. Low frequencies face high reactance through the capacitor, so they are fed out through the output.

The chief specification for a filter is the cut-off frequency. This is the point where the filter stops passing the signals, and starts blocking them. This is not an abrupt switch, but a gradual slope, as illustrated in Fig. 6-27.

The cut-off frequency for the simple passive low-pass filter is:

$$F = 159000/RC$$

The low-pass filter is one of four basic types. If we reverse the position of the components in the low-pass filter, as shown in Fig. 6-28, we get a high-pass filter that behaves in the opposite manner, but the cut-off frequency remains the same.

A somewhat more complex type of filter is the band-pass filter. In this type of circuit only a specific band of frequencies is passed. Anything above or below this band is blocked. The response of a typical band-pass filter is illustrated in Fig. 6-29.

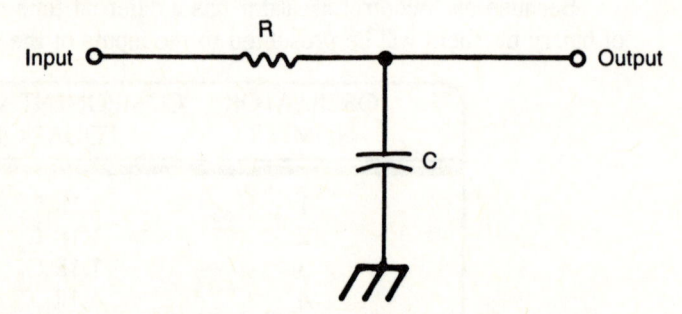

Fig. 6-26. A passive low-pass filter made from a resistor and a capacitor.

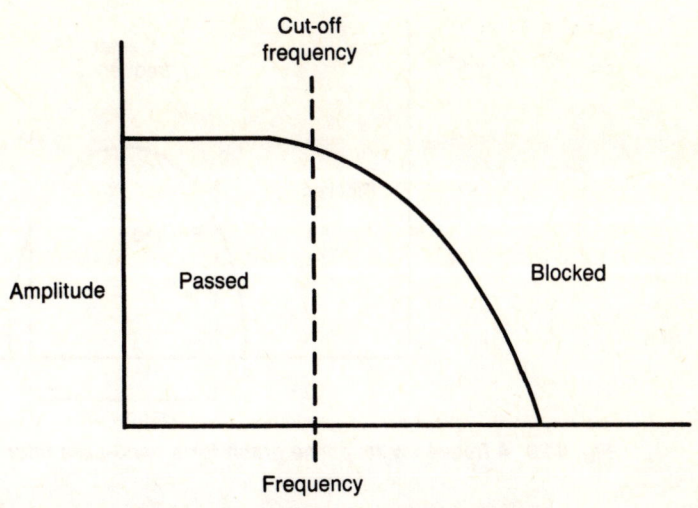

Fig. 6-27. The cut-off slope of a simple filter is gradual, not sharp.

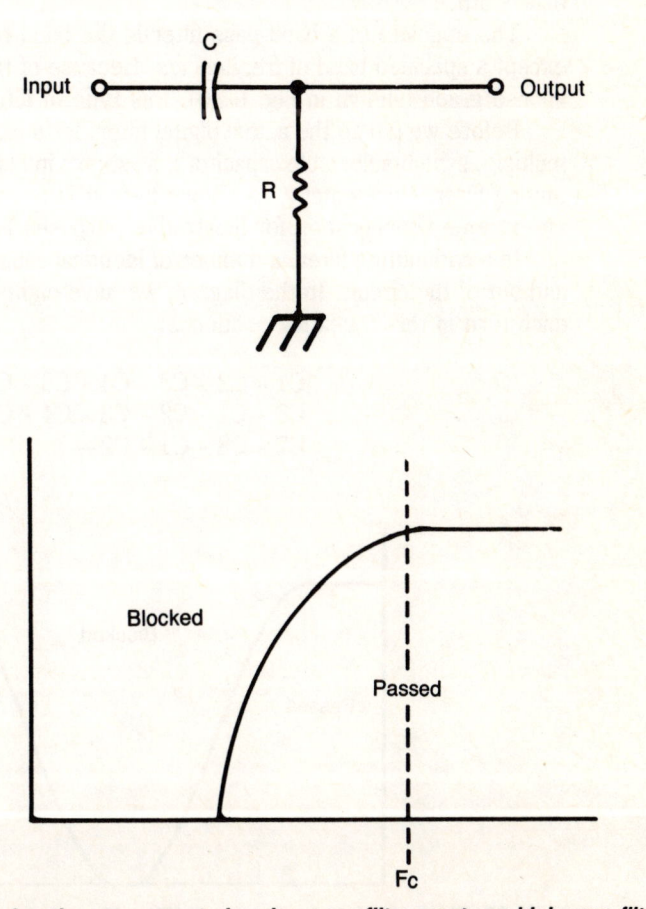

Fig. 6-28. Reversing the components in a low-pass filter creates a high-pass filter.

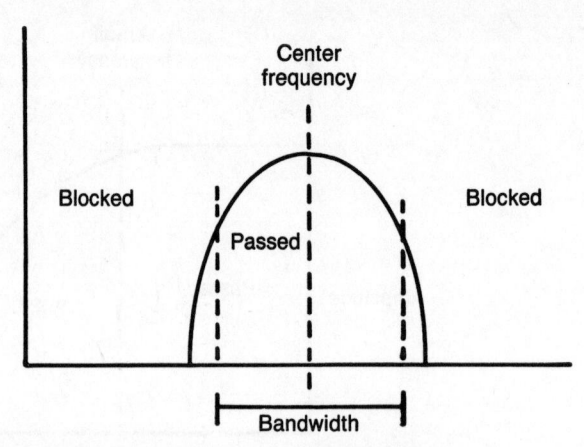

Fig. 6-29. A frequency response graph for a band-pass filter.

A band-pass filter has two key specifications—the center frequency, and the bandwidth.

The opposite of a band-pass filter is the band-reject filter that passes everything except a specified band of frequencies. Because of the appearance of its frequency response graph (shown in Fig. 6-30), this type of filter is often called a notch filter.

Before we get to the actual digital filter, let's expand the basic low-pass filter with multiple, switch-selectable capacitors, as shown in Fig. 6-31. This circuit is called a commuting filter. (Incidentally, as shown here it is <u>not</u> a practical circuit. This diagram is an extreme simplification for illustrative purposes.)

In a commuting filter, a number of identical capacitors are switched sequentially in and out of the circuit. In the diagram we have eight capacitors, so the switch grounds each turn in this repeating sequence:

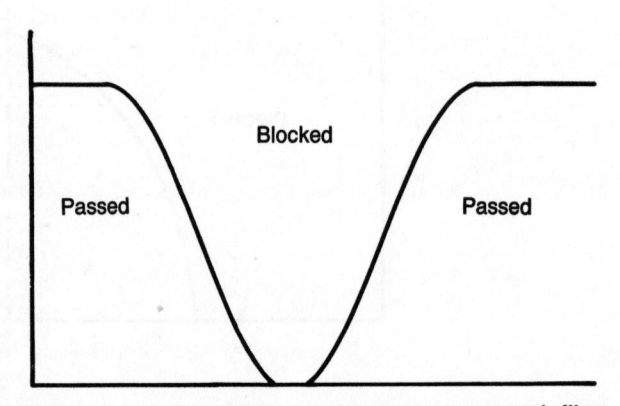

C1 - C2 - C3 - C4 - C5 - C6 - C7 -
C8 - C1 - C2 - C3 - C4 - C5 - C6 -
C7 - C8 - C1 - C2—

Fig. 6-30. The opposite of a band-pass filter is a band-reject, or notch filter.

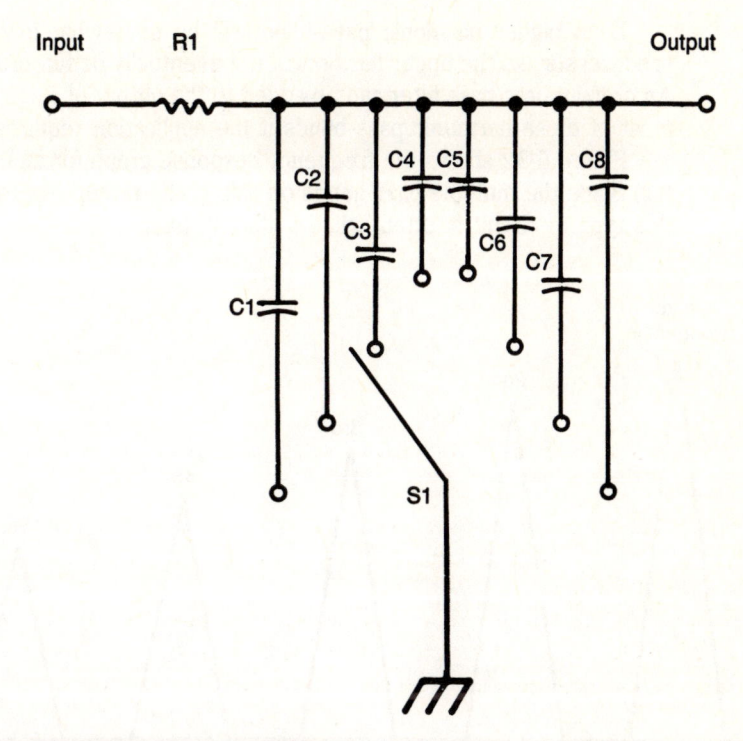

Fig. 6-31. A simplified commuting filter.

This capacitor switching action changes the operation of the filter from a low-pass to a band-pass type. The original cut-off frequency equation is no longer relevant. The equation to determine the center frequency of the pass-band is:

F = 1/2nRC

where *n* is the number of capacitors or switch positions. Remember that all of the capacitors have equal values (C) allowing this simple equation to be used.

Actually, unlike ordinary band-pass filters, a commuting filter actually has a number of pass-bands. Harmonics (integer multiples) of the center frequency will also be allowed to pass through the commuting filter, albeit, with increasing attenuation. For instance, if the center frequency is 1000 Hz, there will be additional pass-bands with center frequencies of:

2000 Hz	(second harmonic)
3000 Hz	(third harmonic)
4000 Hz	(fourth harmonic)
5000 Hz	(fifth harmonic)
6000 Hz	(sixth harmonic)

and so forth.

Each higher harmonic pass-band will be somewhat lower in amplitude than its predecessor, so the upper harmonics will eventually be filtered out of the output signal. An ordinary low-pass filter can be added to the output of a commuting filter to eliminate most of these harmonic pass-bands if the application requires it.

Figure 6-32 shows the frequency response graph for an unmodified commuting filter. Since the multiple pass-bands on this graph rather resemble the teeth of a comb, this type of circuit is often called a "comb filter".

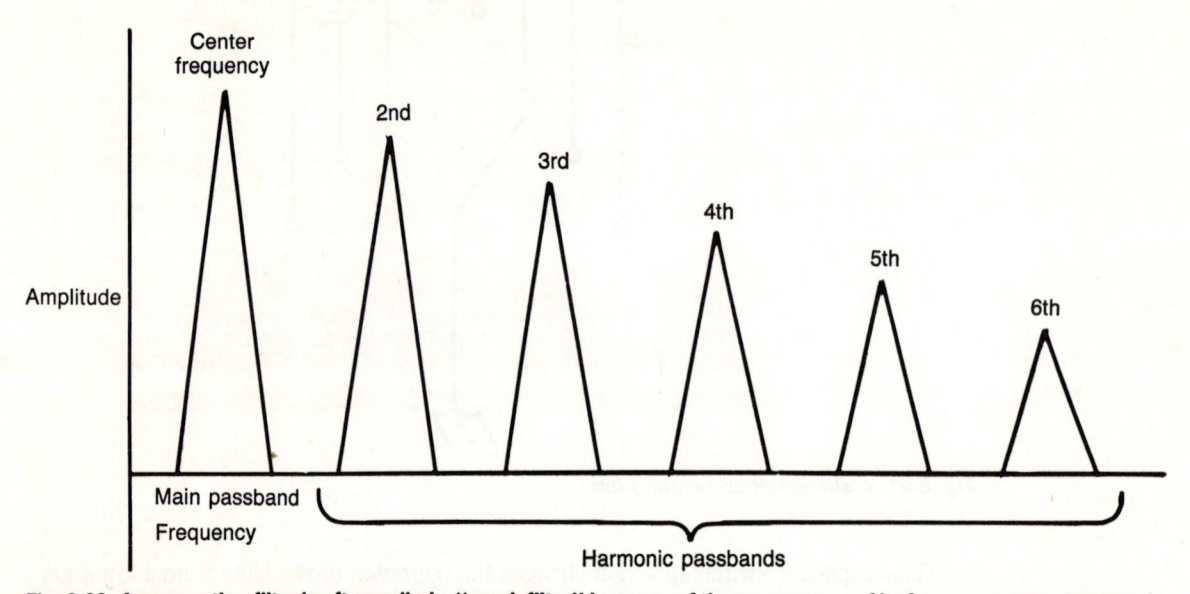

Fig. 6-32. A commuting filter is often called a "comb filter" because of the appearance of its frequency response graph.

The commuting filter circuit shown back in Fig. 6-31 is not functional. There is no way to move the mechanical switch at a fast enough rate. Here is where digital circuitry comes into play.

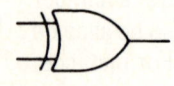

Project 30: Digital Filter

We can use a digital circuit to perform the capacitor switching action. A suitable circuit is shown in Fig. 6-33. Notice that two inputs are required for this circuit. One is a digital clock signal to drive the counter, and the other is the signal to be filtered, this can be any analog signal.

The capacitor values in this circuit are fairly critical. Components with no more than a 10 percent tolerance rating should be used in this application.

A typical parts list for this project is given in Fig. 6-33. Feel free to experiment with other resistor and capacitor values, but remember that all of the filter capacitors must have equal values for the circuit to operate properly.

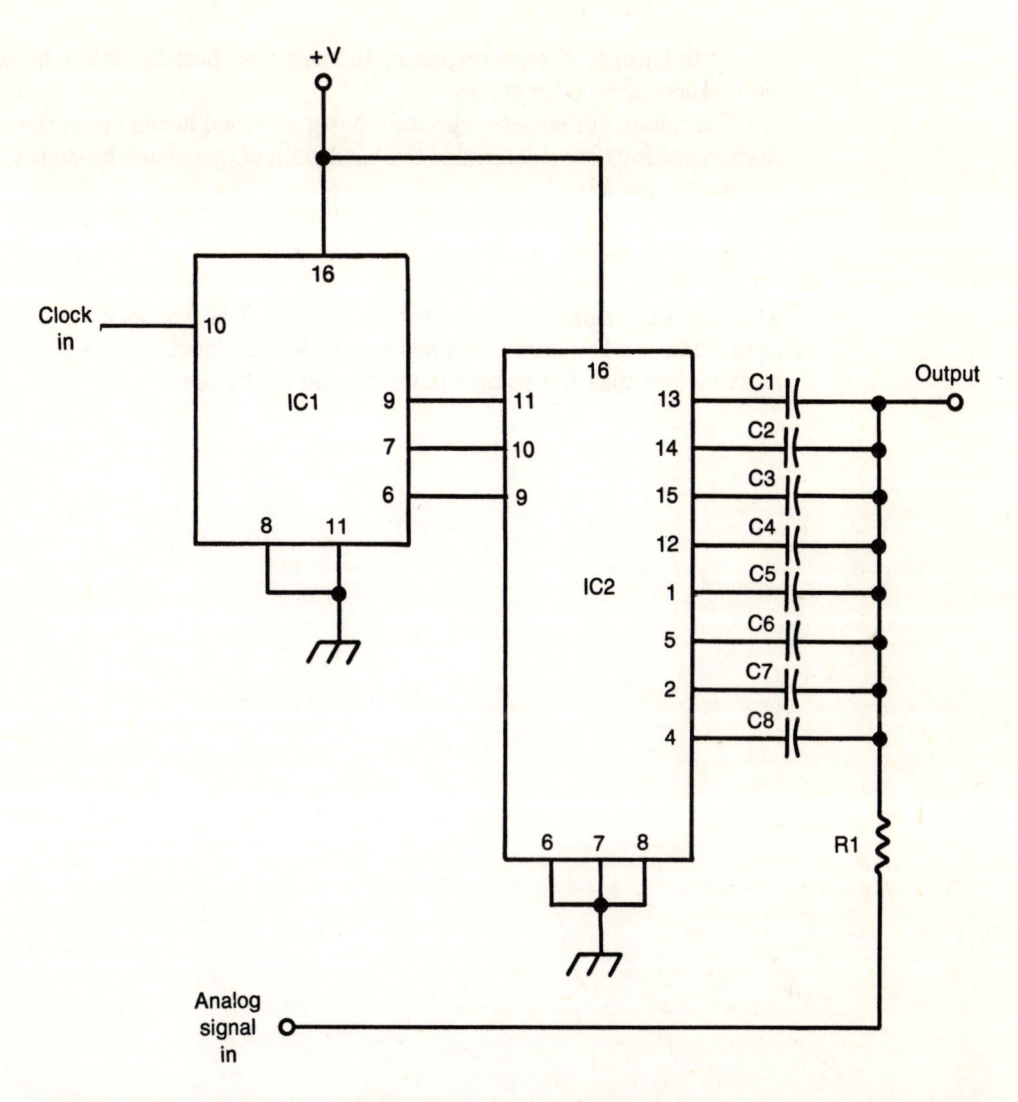

#30 PROJECT PARTS LIST	
COMPONENT	**DESCRIPTION**
IC1	CD4040 BCD ripple counter
IC2	CD4051 BCD-to-decimal decoder
R1	1kΩ resistor *
C1 - C8	0.01 µF capacitor *

* experiment with other component values.
All capacitors should be identical.

Fig. 6-33. A practical digital filter circuit.

More linear frequency response, and narrower pass-bands can be achieved by adding additional decoder stages.

The number of stages (capacitors being switched in and out of the circuit) and the clock signal frequency determine the bandwidth of the center-frequency pass-band, according to this formula:

$$F = X/n$$

where n is the number of capacitor stages, and X is the clock frequency. Almost any input signal can be filtered with this circuit, but the input signal's peak-to-peak voltage <u>must</u> be less than the voltage powering the digital ICs.

7
COUNTER CIRCUITS

A rather obvious application for digital devices is counting. There are more possible variations than you might suspect. In this chapter we will present projects for just a few of the many types of counter circuits. All of these projects could be used as part of a larger system.

Human beings are generally used to dealing with numbers using the ten digits of the familiar decimal system (0, 1, 2, 3, 4, 5, 6, 7, 8, and 9). Most of us are not really accustomed to reading a bunch of dots that are either on (lit) or off (dark), with the position determining their value. The binary (two digit) numbering system is ideal for electronic circuits, but people find it extremely awkward at best.

One-out-of-X sequential counters are an improvement over direct binary outputs, but they are still far from ideal.

When a person must read an output value, it would be preferable to have that output presented directly in decimal form. If the output is in any other form you must translate the value in your head. Why not let the electronic circuitry do the work?

Hooking up a binary counter with a BCD (Binary-Coded-Decimal) decoder and a seven-segment display driver, as illustrated in Fig. 7-1, will accomplish these ends quite nicely.

BCD is a compromise coding system somewhere between true binary and true decimal numbering. Four BCD digits are used to represent each decimal digit. A BCD digit, like a binary digit is either a 1 or 0. In fact, the BCD values are equal to binary values shown in Table 7-1:

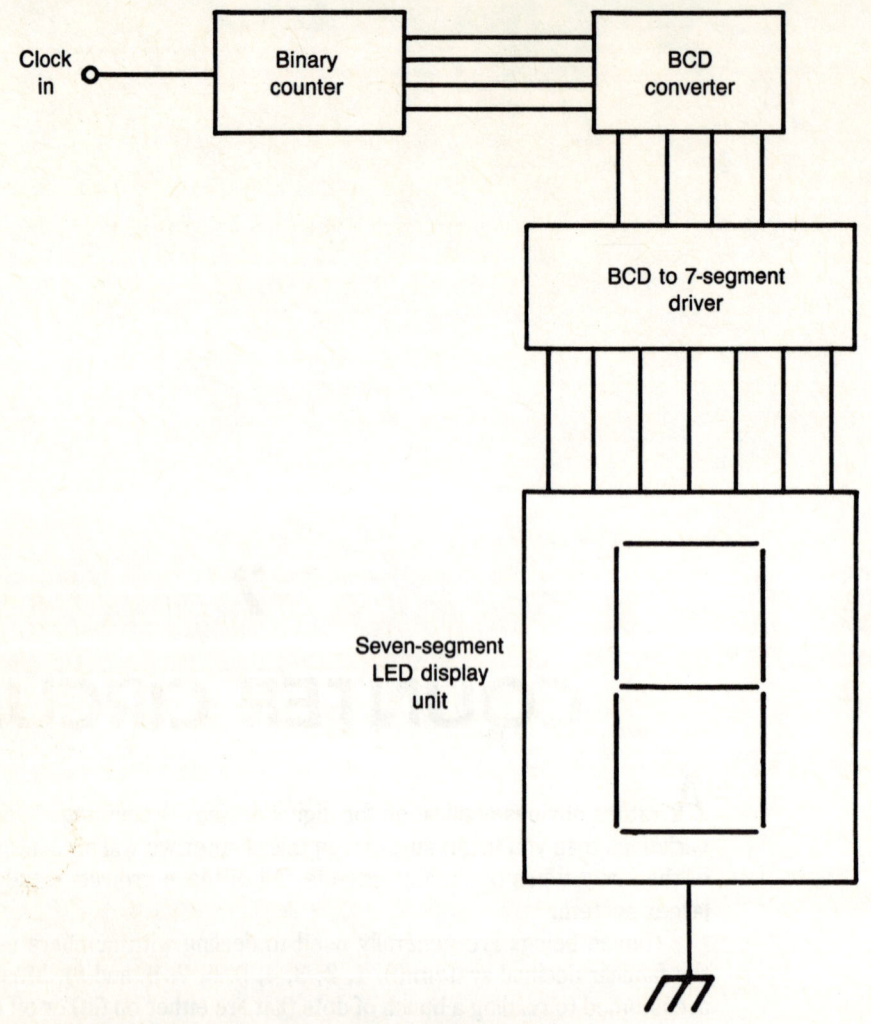

Fig. 7-1. A block diagram for a single digit decimal counter.

Table 7-1.

BCD	DECIMAL	BINARY
0000	0	0000
0001	1	0001
0010	2	0010
0011	3	0011
0100	4	0100
0101	5	0101
0110	6	0110
0111	7	0111
1000	8	1000
1001	9	1001

The difference between the BCD and binary systems is evident when we represent values past nine. Nine is the highest digit in the decimal system. To represent a number larger than nine, we must add another digit. Similarly, in BCD we add another four-bit digit group, while the binary equivalent continues uninterrupted as in Table 7-2:

BCD	DECIMAL	BINARY
0001 0000	10	1010
0001 0001	11	1011
0001 0010	12	1100
0001 0011	13	1101
0001 0100	14	1110
0001 0101	15	1111
0001 0110	16	10000
0001 0111	17	10001
0001 1000	18	10010
0001 1001	19	10011
0010 0000	20	10100
0010 0001	21	10101
0010 0010	22	10110
0010 0011	23	10111

Table 7-2.

In the BCD system, certain four-bit combinations are never used. These are called disallowed states:

1010; 1011; 1100; 1101; 1110; 1111

These six disallowed bit combinations do not correspond to a decimal digit.

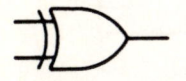

Project 31: Decimal Output Counter

Converting binary values to BCD form is rather tricky, but fortunately, there are many dedicated ICs that will do the job for you. For example, the CD4518 is a dual BCD counter. It behaves like a binary counter (actually two counters in a single package), but the output is in BCD form.

A decimal digit counter circuit is shown in Fig. 7-2 accompanied by the parts list for this project.

Besides the CD4518 BCD counter, this circuit uses the CD4511 seven segment display driver IC.

A seven-segment display is made up of seven individually segmented rectangular LEDs arranged in a figure-eight pattern. One end of all the LEDs is common. A common-cathode display unit is used in this project, but common-anode units are also available.

Fig. 7-2. A single digit decimal counter.

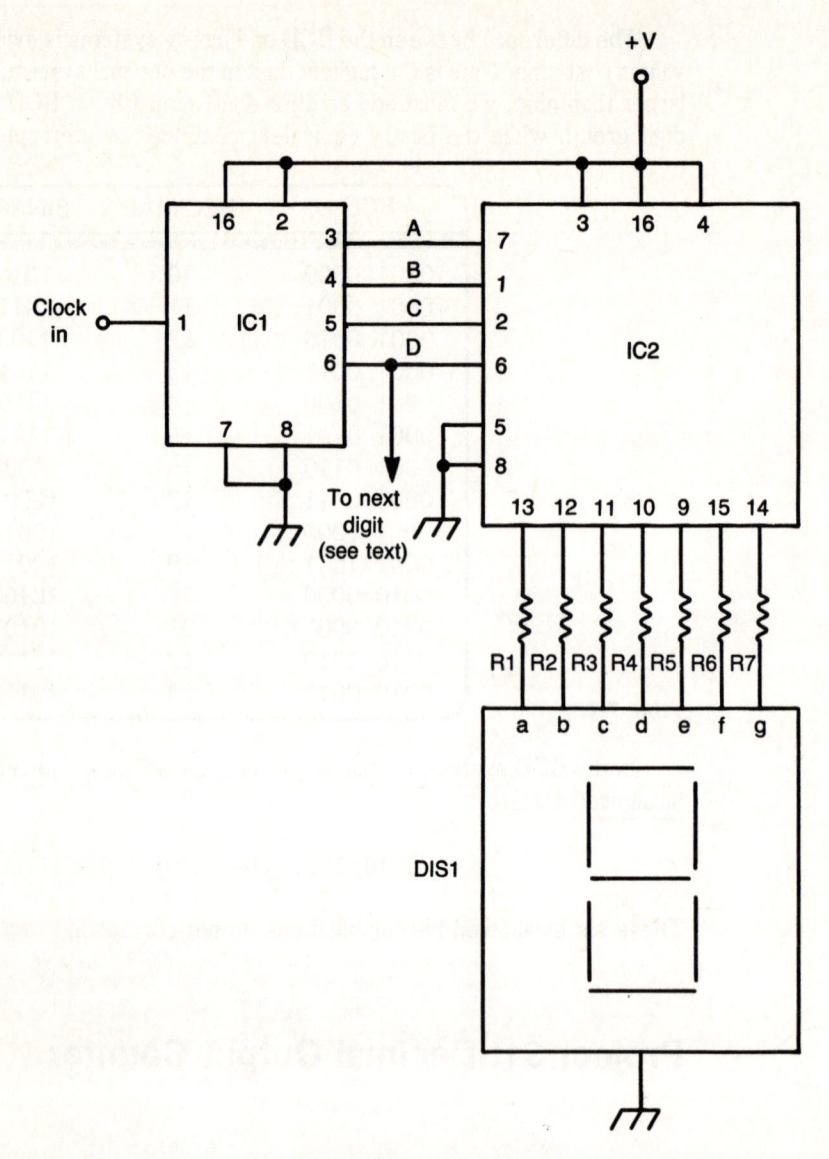

#31 PROJECT PARTS LIST	
COMPONENT	**DESCRIPTION**
IC1	CD4518
IC2	CD4511
DIS1	SEVEN SEGMENT LED DISPLAY UNIT (Common Cathode)
R1 - R7	330Ω resistor

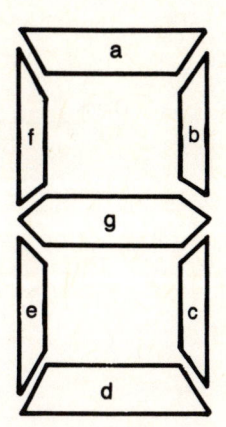

Fig. 7-3. *Any decimal digit can be formed from a 7-segment display unit.*

The seven segments are identified by lower case letters from a to g, as shown in Fig. 7-3. By selectively lighting the appropriate LED segments, any decimal digit from 0 to 9 can be displayed. The CD4511 converts the BCD value into signals to the appropriate LED segments demonstrated by Table 7-3.

It is very easy to expand this circuit for more than a single digit. The technique is illustrated in the block diagram of Fig. 7-4. Simply use the D (most significant digit) line (pin #6 of IC1) as the clock input for a second identical stage. When this pin goes from HIGH (1) to LOW (0) (after a count of nine, changing to a count of zero), the second counter stage is triggered, incrementing the second stage count by one. This same approach can be used for any desired number of display digits.

Table 7-3.

BCD INPUTS	a	b	c	d	e	f	g	DECIMAL VALUE
0000	0	0	0	0	0	0	-	0
0001	-	0	0	-	-	-	-	1
0010	0	0	-	0	0	-	0	2
0011	0	0	0	0	-	-	0	3
0100	-	0	0	-	-	0	0	4
0101	0	-	0	0	-	0	0	5
0110	0	-	0	0	0	0	0	6
0111	0	0	0	-	-	-	-	7
1000	0	0	0	0	0	0	0	8
1001	0	0	0	-	-	0	0	9
1010	DISALLOWED STATE							?
1011	DISALLOWED STATE							?
1100	DISALLOWED STATE							?
1101	DISALLOWED STATE							?
1110	DISALLOWED STATE							?
1111	DISALLOWED STATE							?

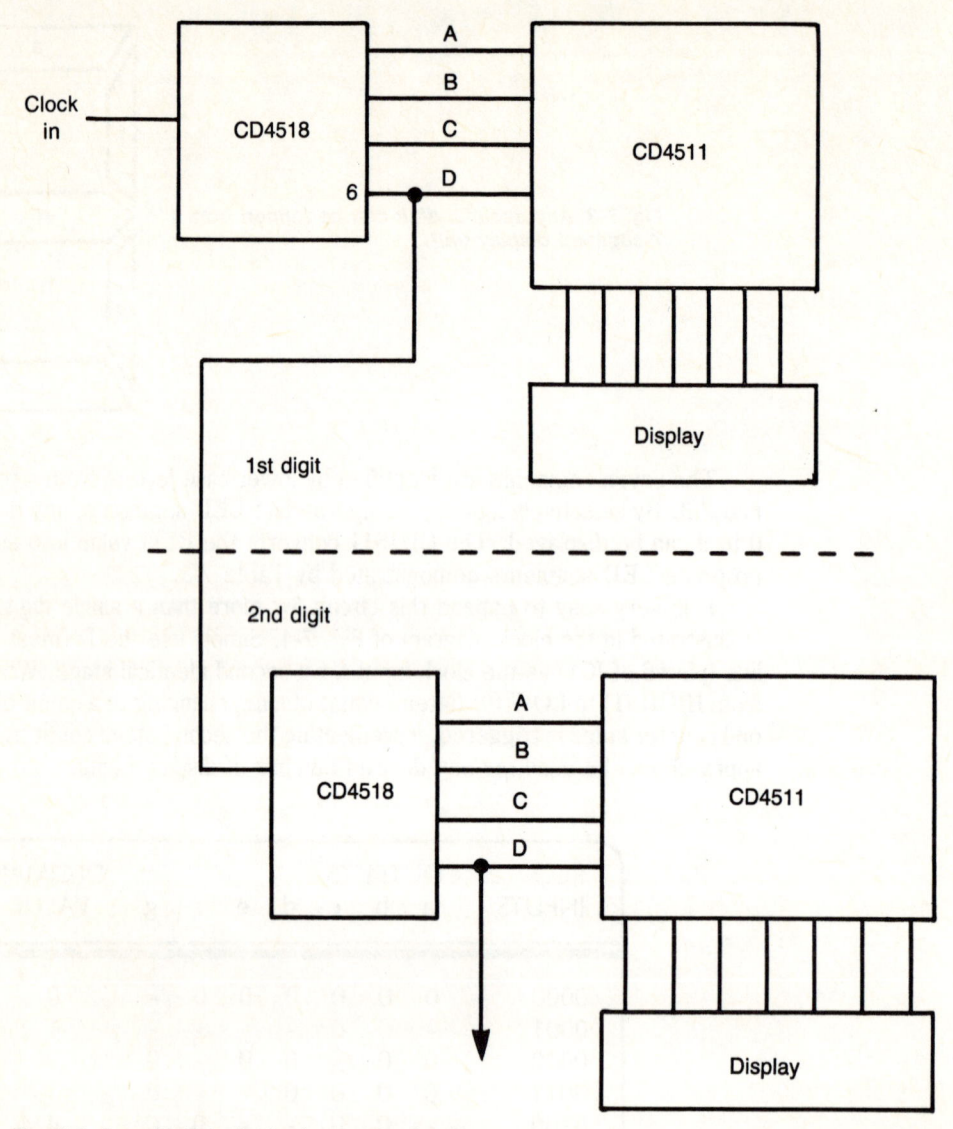

Fig. 7-4. A multi-digit decimal counter circuit.

MULTI-DIGIT COUNTERS

It is fairly easy to create a simple, but entirely functional counter circuit by stringing a series of flip-flop stages together, so that each stage triggers the next, as illustrated in Fig. 7-5.

For many purposes this type of counter is perfectly adequate. However, it is not ideal for applications involving a direct read-out. This type of counter counts in the binary number system, which has only two digits (0 and 1).

People are far more comfortable with the decimal system with its ten digits (0 through

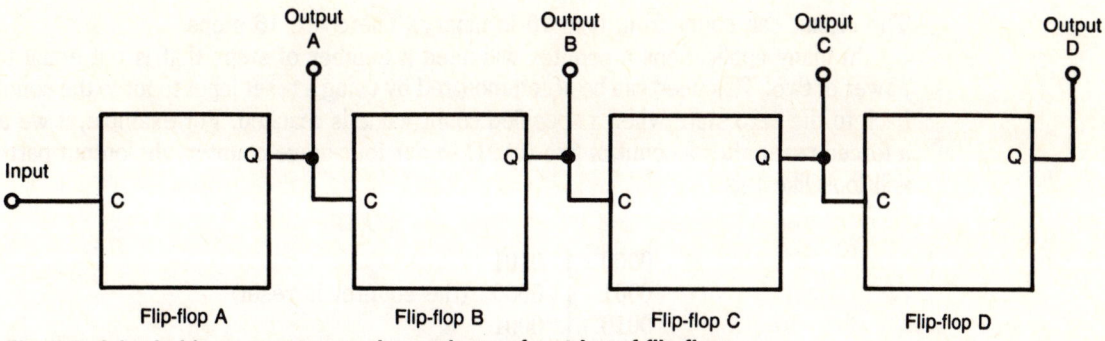

Fig. 7-5. A basic binary counter can be made up of a string of flip-flops.

9). As a comparison, consider the output pattern for a four-stage binary counter shown in Table 7-4.

The maximum count is equal to $2^N - 1$, where N is the number of flip-flop stages. In the four-stage counter of our example, the maximum count is equal to:

$$2^N - 1 = 2^4 - 1$$
$$= 16 - 1$$
$$= 15$$

BINARY OUTPUT	DECIMAL EQUIVALENT
0000	0
0001	1
0010	2
0011	3
0100	4
0101	5
0110	6
0111	7
1000	8
1001	9
1010	10
1011	11
1100	12
1101	13
1110	14
1111	15
0000	0 (16)

Table 7-4.

(The counter resets itself and the pattern repeats.)

The circuit can count from 0 to 15 in binary. There are 16 steps.

In many applications a counter will need a number of steps that is not equal to a power of two. This need can be accommodated by using a reset input to force the counter back to the zero state when a specified count value is reached. For example, if we add a forced reset after a count of five (0101) to our four-stage counter, the output pattern will look like this:

0000	0101	
0001	0000	(the counter is reset)
0010	0001	
0011	0010	
0100		

The number of count steps in the output pattern (including zero) is called the modulo of the counter. In our example, the unmodified four-stage counter had a modulo of sixteen. By adding a forced reset after a count of five (0101), we created a modulo-six counter. (Notice that in this case, the fourth stage isn't used at all.) A counter with any modulo can easily be designed in this manner.

It is rarely necessary to use separate flip-flops. Dedicated binary counter ICs are readily available in a number of variations.

Binary counters are easy to design and are enormously practical in electronic systems. Since each digit can take on one of just two possible values, numbers can easily be represented by simple ON/OFF or HIGH/LOW signals. But since people normally have ten fingers and ten toes, we all learned to count in the decimal system. The ten digit system seems natural to us, while the two digit binary system is strange and awkward, especially when large values are involved. For example, what is the value of this eight digit binary number?

$$10011100$$

When a counter's output is to be displayed and read by a human operator, a simple binary output is obviously undesirable.

One common solution is to adapt the binary values to a one-in-ten output. In this system, we have ten separate outputs (one for each digit from 0 to 9). One is activated at any time, while the other nine are deactivated. The active output indicates the current value. This system gives an output that is much clearer and easier to comprehend by the operator.

The required gating to create a one-in-ten counter is fairly complicated. Fortunately, we don't need to be bothered with the details. One-in-ten decimal counters are readily available in IC form. One such device is the CD4017, which is illustrated in Fig. 7-6. Notice that the CD4017 has ten separate outputs numbered from 0 to 9. On any specific count, only one of these outputs will be HIGH, and the others will all be LOW.

The CD4017 can be operated in two different modes—Count-and-Halt, and Continuous. In either mode, the counter can be forcibly reset at any desired point in the count sequence.

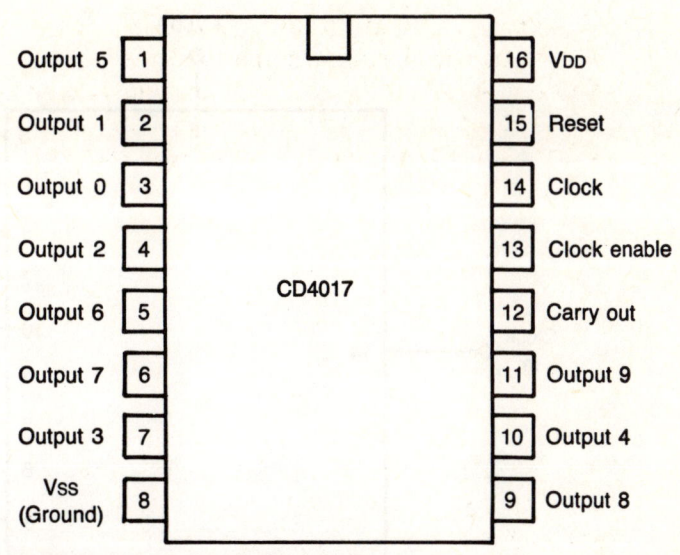

Fig. 7-6. The CD4017 is a one chip digital counter.

By grounding pin #15 of the CD4017, and connecting pin #13 to one of the outputs, the counter will count from 0 to that output's value, and then stop. Figure 7-7 shows an example of a circuit using the CD4017 in the Count-and-Halt mode. In this circuit the maximum count is seven. After the maximum count is reached, further input pulses will be ignored by the counter. The circuit can be reset for another count cycle by temporarily disconnecting pin #15, and momentarily connecting it to a positive voltage source. Generally the V_{DD} supply voltage will be used for this purpose.

Figure 7-8 illustrates how the CD4017 can be used in the Continuous mode. Here the connections of pin #13 and pin #15 are simply reversed from the Count-and-Halt

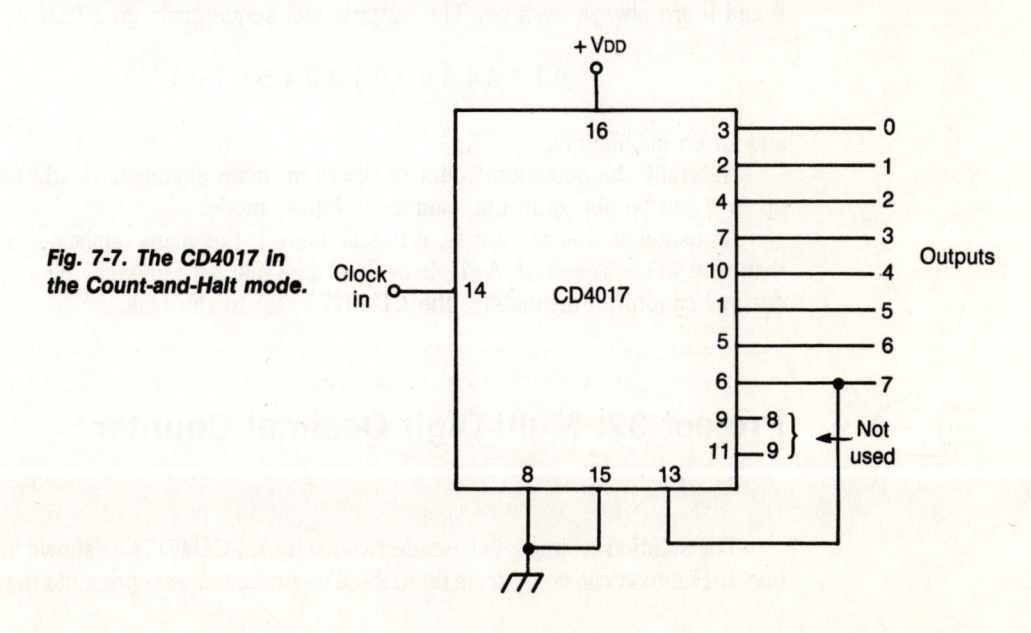

Fig. 7-7. The CD4017 in the Count-and-Halt mode.

125

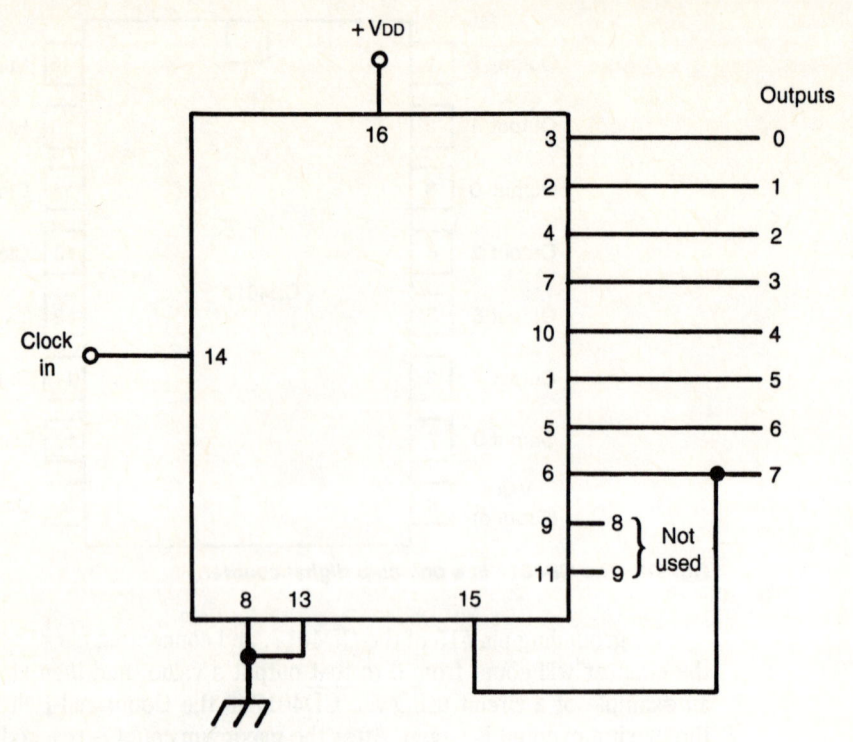

Fig. 7-8. The CD4017 in the Continuous mode.

mode. In the Continuous mode, pin #13 is grounded and pin #15 is connected to the output representing the maximum desired count. In the figure a maximum of 7 is used again. The outputs will count from 0 to 7 then jump back to 0, and start over. Outputs 8 and 9 are always inactive. The outputs will sequentially go HIGH in this order:

0 1 2 3 4 5 6 7 0 1 2 3 4 5 6 7 0 1 2 3 . . .

and so on indefinitely.

Naturally the maximum count of seven in these examples is arbitrary. Any count up to 9 can be set up in this manner in either mode.

As useful as this counter is, it is still limited. For many applications counts higher than nine will be required. A single decimal digit just isn't enough. We need a multi-digit decimal counter. Fortunately, the CD4017 is up to the task.

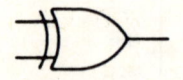

Project 32: Multi-Digit Decimal Counter

The solution is simply to cascade two (or more) CD4017s as shown in Fig. 7-9. This two 4017 circuit can count from 00 to 99. The first counter represents the one's column,

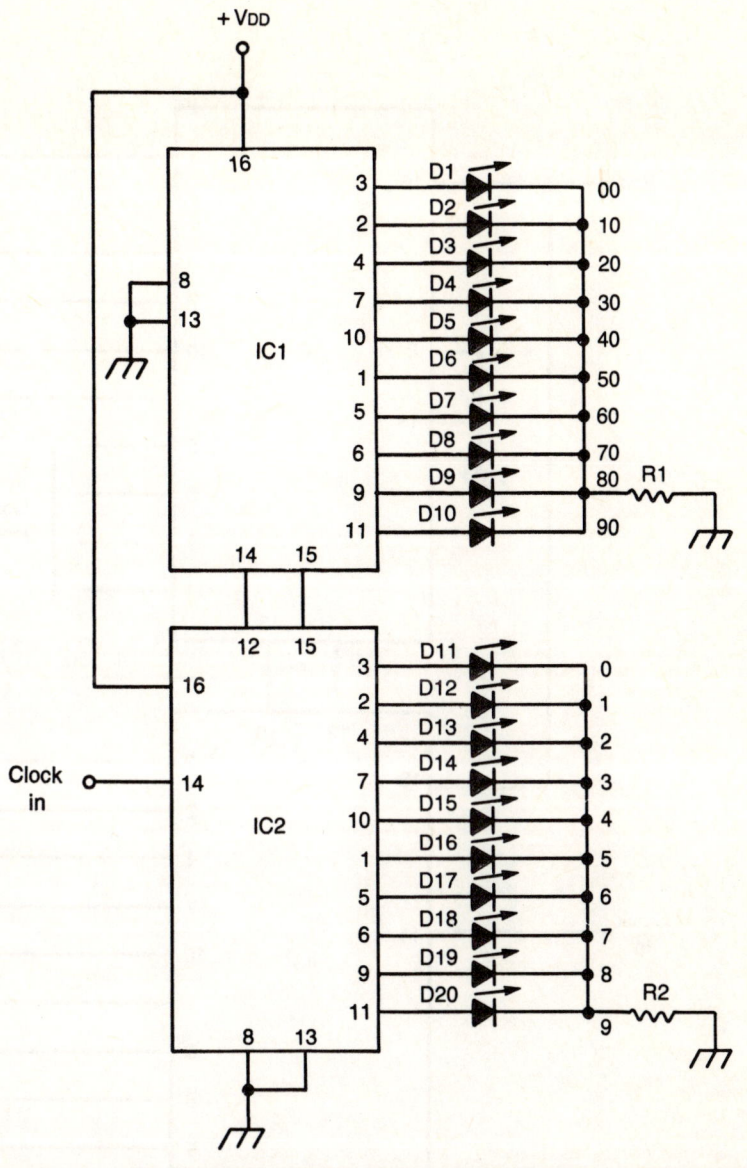

Fig. 7-9. CD4017s can be cascaded for multi-digit operation.

#32 PROJECT PARTS LIST	
COMPONENT	**DESCRIPTION**
IC1, IC2	CD4017 counter
D1 - D20	LED
R1, R2	470Ω resistor

127

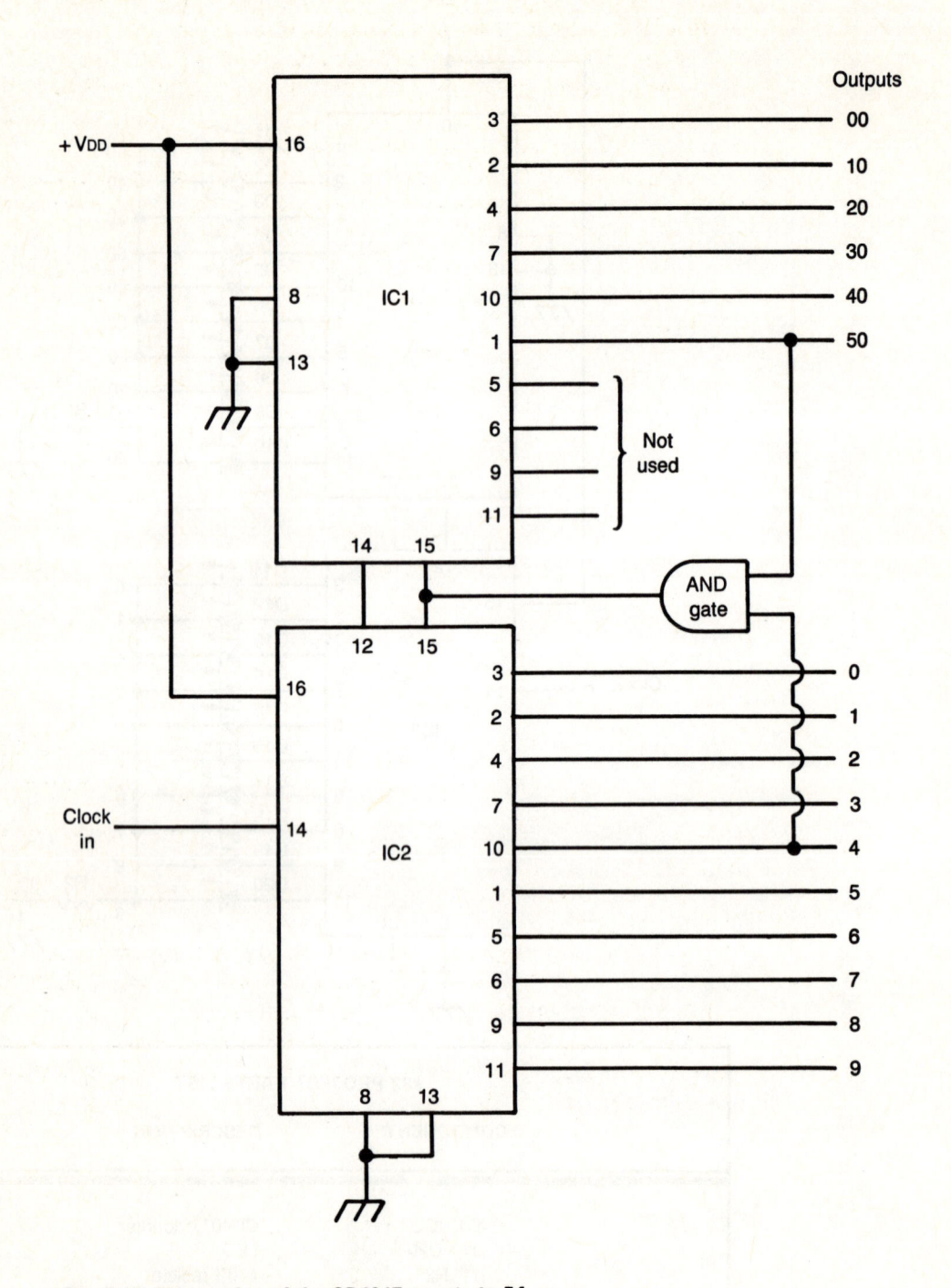

Fig. 7-10. This version of the CD4017 counts to 54.

128

and the second counter handles the tens column. If a third stage is added, it will add the hundreds column for a count range from 000 to 999.

A parts list for this project is given in the figure. As you can see, not many parts are required beyond the CD4017s. For demonstration purposes, LEDs are used to indicate the output state. The LEDs, and the current-dropping resistors can be eliminated if they are not needed in your individual application.

Since only one LED will ever be lit in either group at any time, the LEDs in each group can share a current-dropping resistor, as shown here. You could use separate resistors for individual LEDs, but I don't see much point to it.

The exact value of the current-dropping resistor is not particularly critical, increasing the resistance will decrease the LEDs' brightness, while decreasing the resistance will cause the LEDs to burn brighter.

If you need a counter with a modulo less than 100, but greater than 10, the circuit can easily be adapted. A two-input gate (usually an AND gate) will be needed because two outputs (one from each CD4017) will be needed to uniquely indicate the desired reset point. The circuit of Fig. 7-10 counts to 54 in the Continuous mode. The multi-digit counter can also be used in the Count-and-Halt mode by reversing the connections to pins #13 and #15.

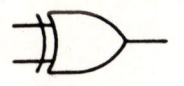

Project 33: Decimal Count-Down Timer

In many applications it is useful to have a timer circuit that counts backwards, indicating the amount of time remaining before the circuit time cycle ends. This type of device could be useful in a darkroom or a kitchen.

The 1 hertz timebase circuit is used to drive the counters, this circuit is presented as Project #38.

The rest of the circuit is broken up into two schematics for convenience. Figure 7-11 is the switching circuit for determining the start value for the count-down timer. Figure 7-12 shows the counter and display circuitry. Although only a single digit is shown here this circuit can be expanded to include any number of stages, simply use the Borrow Out pin (#13) of IC2 to serve as the input signal (pin #4) for the next higher stage.

The switching circuit of Fig. 7-11 is simple enough. The operator can enter the desired starting value from 1 to 9 in the proper BCD format. As a reminder, here are the BCD equivalents for each digit:

1	0001	6	0110
2	0010	7	0111
3	0011	8	1000
4	0100	9	1001
5	0101		

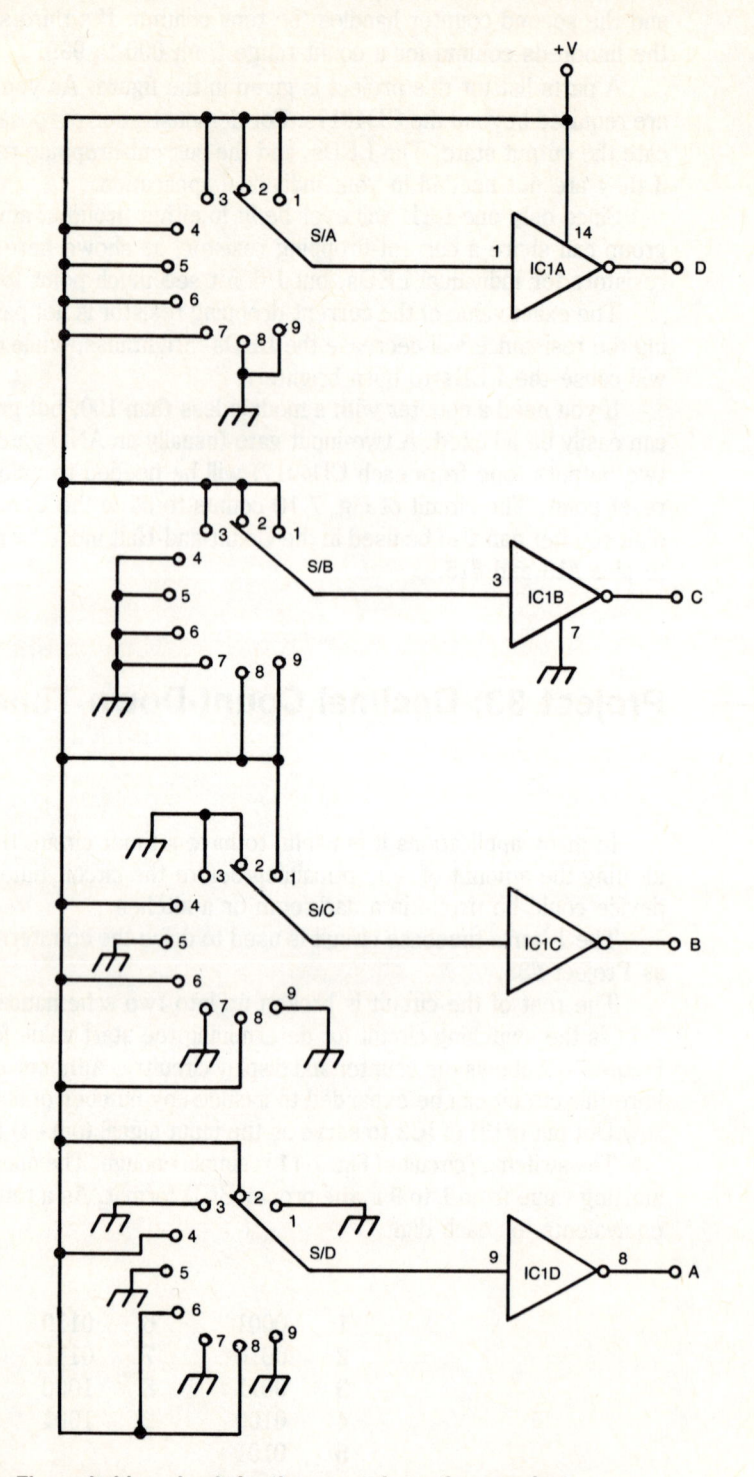

Fig. 7-11. The switching circuit for the count-down timer project.

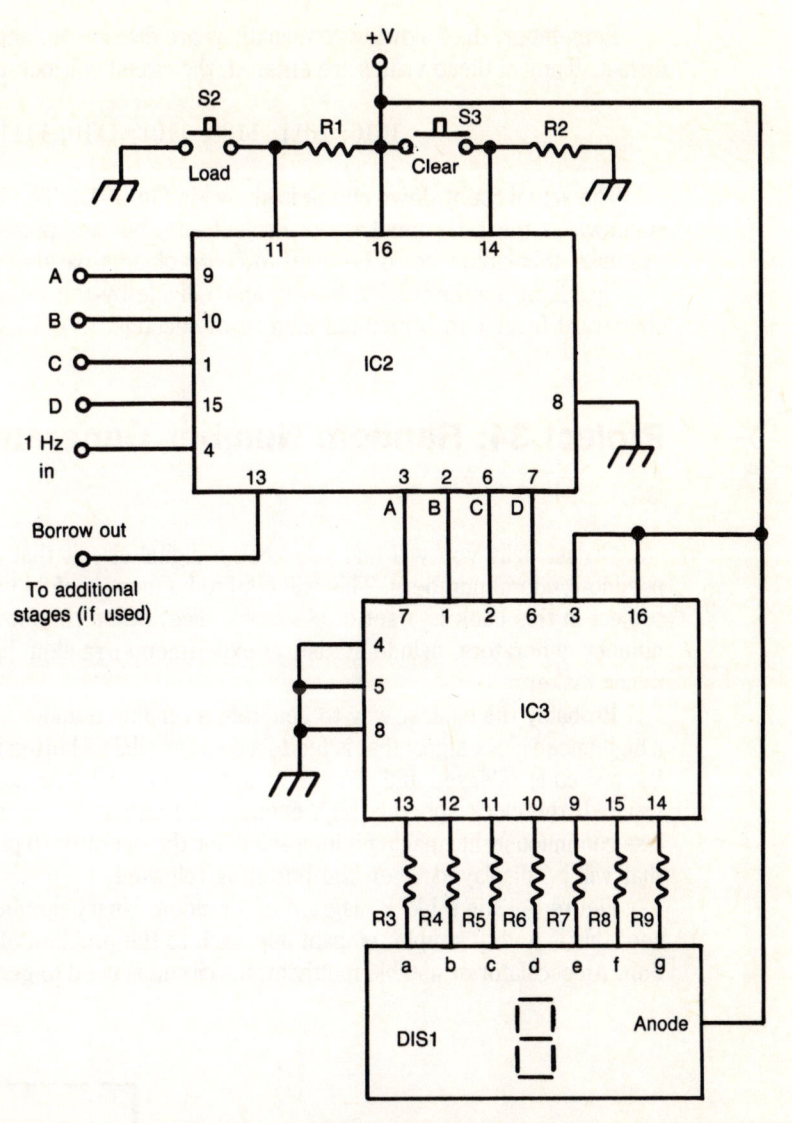

#33 PROJECT PARTS LIST	
COMPONENT	**DESCRIPTION**
IC1	CD4009 hex inverter
IC2	74C193 up/down counter
IC3	74C46 BCD to 7-segment decoder
DIS1	7-segment LED display, common anode
R1, R2	1kΩ resistor
R3 - R9	330Ω resistor
S1	4 pole, 9 throw rotary switch
S2, S3	SPST push-button (Normally Open)

Fig. 7-12. The actual timer section of the count-down timer project.

131

Remember, the following combinations are disallowed, and meaningless in the BCD format. If any of these values are entered, the circuit will lock-up, or perform erratically:

1010; 1011; 1100; 1101; 1110; 1111

The actual count-down circuit is shown in Fig. 7-12. The input signal to be counted is shown as the 1 Hz timebase of Project #38, but any pulse signal can be used. For example, this circuit could be used to count objects passing a detector of some sort.

By adding another divide-by-six, and a divide-by-ten counter, the circuit can count downward from 1 to 9 minutes instead of seconds.

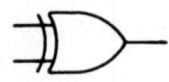

Project 34: Random Number Generator

Occasionally you will find you need a digital circuit that can generate random (or pseudo-random) numbers. This will tend to be most useful in games. For example, one project in this book is a set of electronic dice, other common applications for random number generators include statistical experiments, random light displays, and random music makers.

Probably the easiest way to generate a random number in a digital circuit is to use a high-speed clock signal that is fed to a counter while a button is being depressed. When the switch is released and allowed to open the circuit, the counter stops at its current value. If the clock speed is high enough, the output LEDs will appear to be more or less continuously lit, making it impossible for the operator to predict the final count value that will be displayed when the button is released.

Figure 7-13 is a block diagram of a random binary number generator. As you can see, this is a very simple, elegant approach to the problem of random number generation. An oscillator or astable multivibrator circuit is used to generate the high-frequency

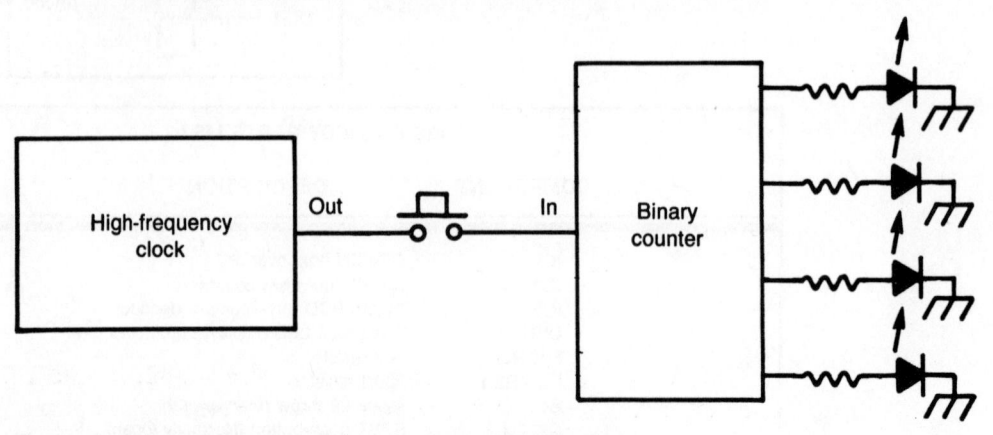

Fig. 7-13. The block diagram of a random number generator.

clock signal. The 555 or 7555 timer would be well suited to this application. The clock rate should be very high in frequency. I'd recommend a clock rate in the 20 kHz to 50 kHz range. The exact frequency is not particularly critical, as long as the capabilities of the counter are not exceeded.

The clock feeds (through the push-button) are four flip-flops, arranged as a four-stage binary counter. When the push-button is closed, the clock signal can get through to the counter. When the button is released, the counter will stop incrementing, and the last count value will be displayed on the output LEDs.

Incidentally, this is one type of digital circuit where a switch debouncing network is never necessary. If the switch bounces, and a few more clock pulses get through to the counter, what difference does it make? In this case we want a randomized output.

Figure 7-14 shows how the same approach can be used to display randomized decimal digits on a seven-segment LED display. The parts list for the project is given in Fig. 7-14.

This circuit works in the same way as the simpler binary version discussed previously. The biggest difference is the four-stage binary counter (flip-flops) is replaced with a CD4518 BCD counter, followed by a CD4511 BCD to 7-segment converter.

While the switch is closed, all seven segments of the display will appear to be constantly lit, so the readout will appear to be a solid 8. (There might be some flicker.) However, once the switch is released the last count value will be displayed.

As with most other counters, additional stages can be cascaded to allow for higher counts. Cascading two of the decimal counters will allow for random numbers from 00 to 99.

There are two convenient ways to cascade decimal number generators. The approach shown in Fig. 7-15 is essentially the same method used for a two (or multi-) digit decimal counter. When counter A passes from 9 to 0, a clock pulse is fed into counter B (tens).

Alternatively, separate clocks could be used to drive the two stages, as illustrated in Fig. 7-16. This creates an even more random effect (especially at lower clock rates). The two clocks should be set at different frequencies that are not harmonically related.

Figure 7-17 shows an approach that might look like it would work, but it doesn't. A single clock is used to drive the two counter stages simultaneously. This is an economical idea, but the results are disappointing. If you think about it for a minute, you should be able to see that both digits will always display the same values (i.e., 55, or 77). Obviously this simply wastes the second display.

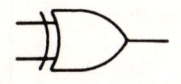

Project 35: Photoelectric Counter

Many light-sensitive electronic components are available today. By mechanically blocking off a light source from a photosensitive device, an object counter, or rotation counter can be easily set up.

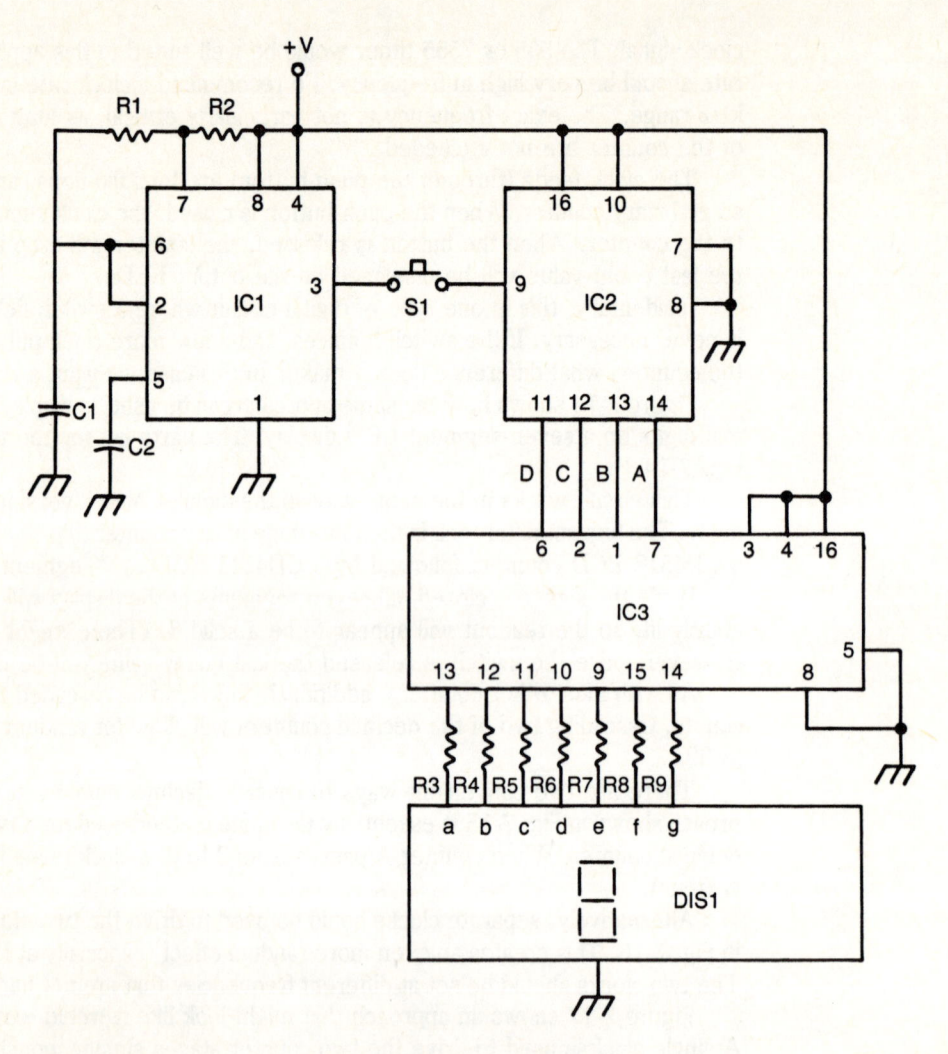

#34 PROJECT PARTS LIST	
COMPONENT	DESCRIPTION
IC1	7555 CMOS timer (a 555 standard timer may be substituted)
IC2	CD4518 BCD counter
IC3	CD4511 BCD to 7-segment decoder
DIS1	Common cathode seven-segment LED display
C1, C2	0.01 µF capacitor
R1, R2	1kΩ resistor
R3 - R9	330Ω resistor
S1	SPST push-button switch (Normally Open)

Fig. 7-14. *The complete random number generator circuit.*

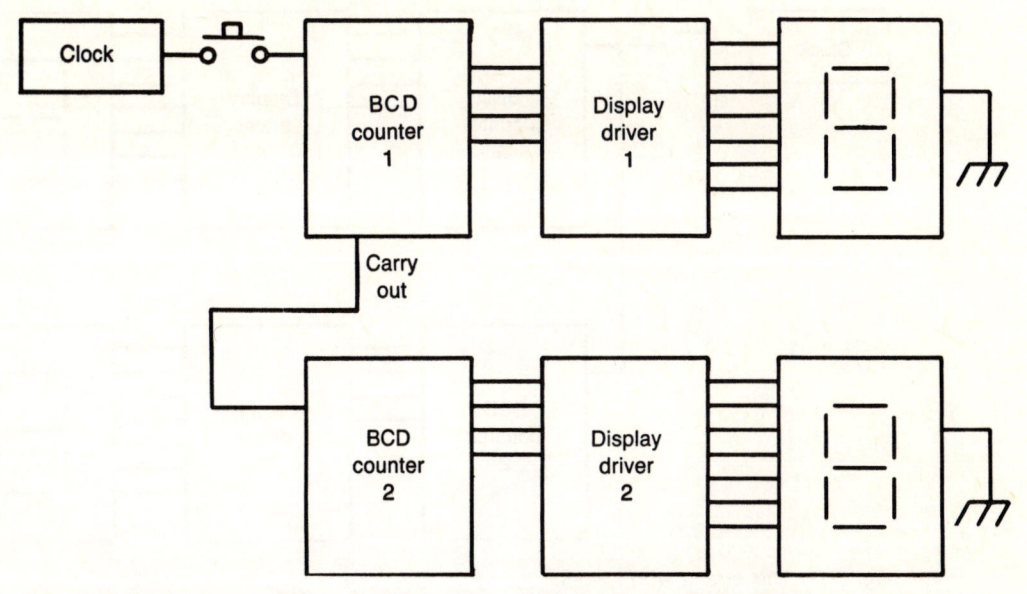

Fig. 7-15. The random number generator can be expanded for multiple digits.

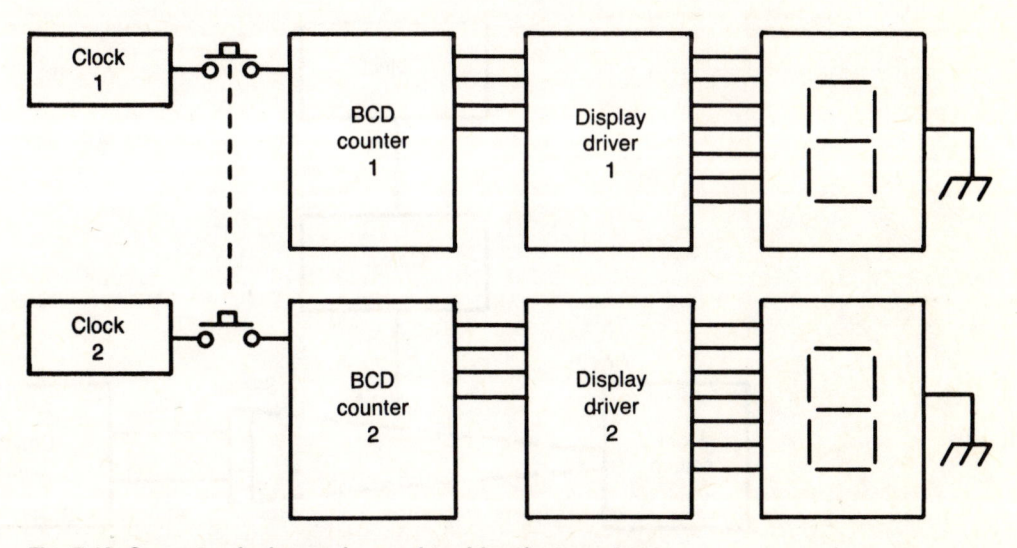

Fig. 7-16. Separate clocks can be used to drive the two stages.

Figure 7-18 illustrates the basic set-up for a simple photoelectric object counter. A light source and the photosensitive sensor are placed across the object path, directly opposite each other. Normally, the light from the source will shine directly on the sensor. But when an object passes between the light source and the sensor, the light beam, as seen by the sensor, is momentarily cut off. The sensor responds to this change in its received light, triggering the counter circuitry. This system can be used to count objects passing by on a conveyor belt, or people passing through a doorway, other applications are also possible.

135

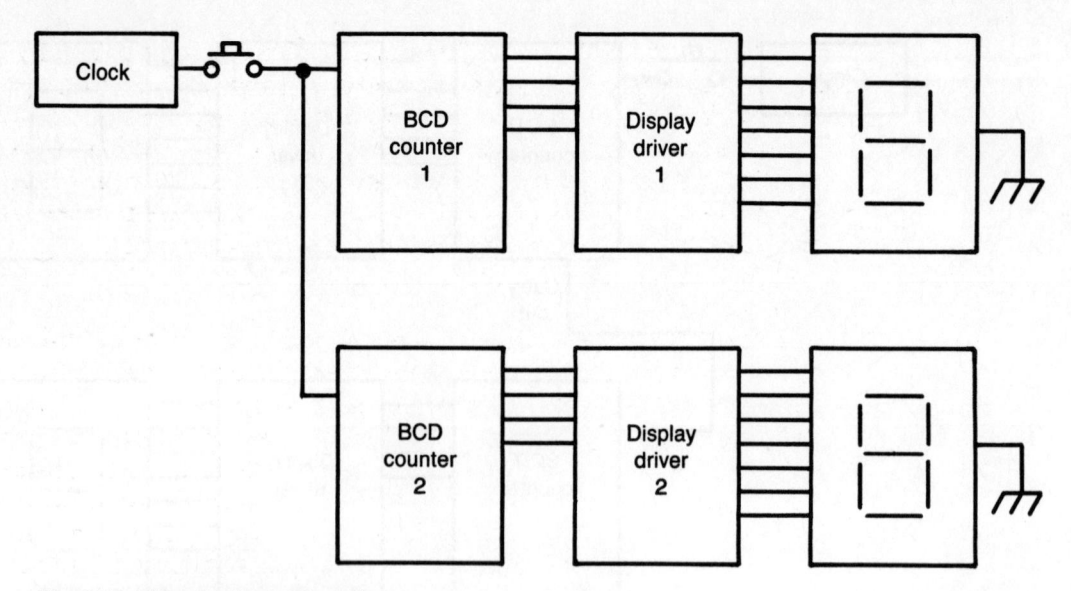

Fig. 7-17. This circuit won't work.

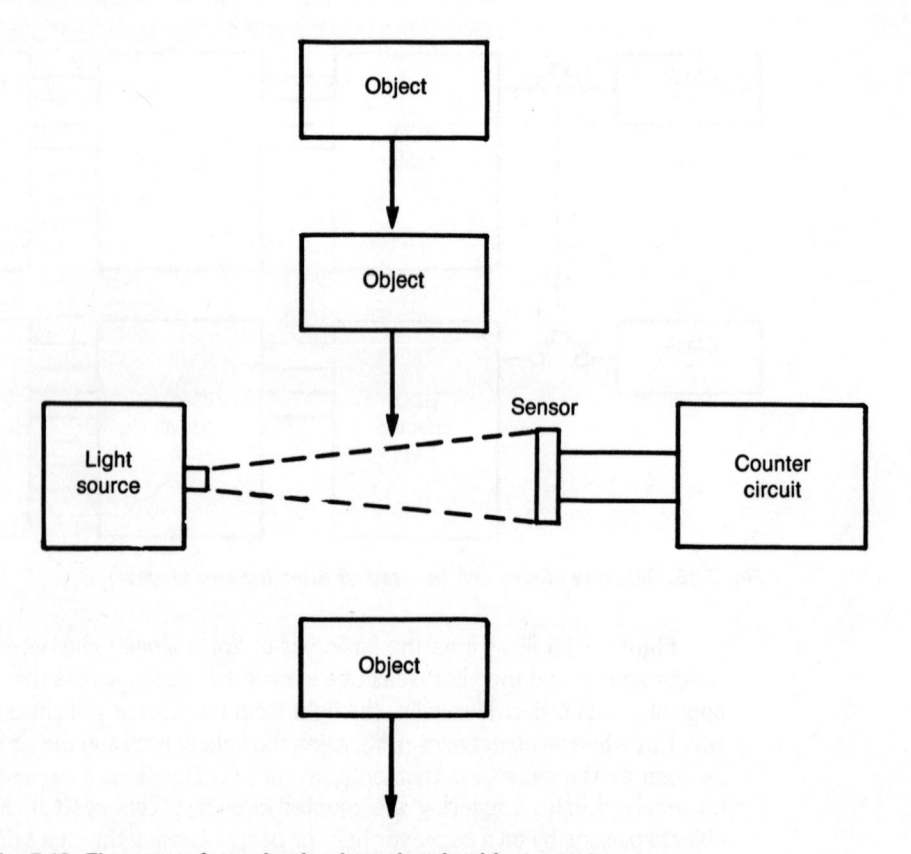

Fig. 7-18. The set-up for a simple photoelectric object counter.

A photoelectric rotation counter works on the same principle of blocking the light source from the sensor to trigger the counter. As shown in Fig. 7-19, a light source and a photoelectric sensor are placed opposite each other. A shaft with an eccentric cam on the end is mounted on the revolving wheel, or whatever rotating object is being monitored. The shaft rotates with the wheel. Once per revolution, the wide end of the cam passes between the light source and the sensor momentarily breaking the light beam. These pulses are counted by a circuit similar to a frequency meter.

A complete photosensitive counter circuit is shown in Fig. 7-20, the parts list for this project is given with it.

Transistor Q1 is a phototransistor used as the light sensor. Notice that there is no electrical connection to the base of this transistor. The light striking its photosensitive surface creates a voltage that functions as the base signal.

IC1 is an op amp (operational amplifier), such as the common and inexpensive 741. The op amp is wired as a comparator. Potentiometer R4 is adjusted during calibration so that breaking the beam of light striking the surface of Q1 will cause the comparator to emit a clean output pulse. This pulse is then used to trigger a monostable multivibrator (IC2) serving as a switch debouncer to clean up any erratic portions of the signal.

IC3 is a timebase oscillator producing a 60 Hz signal. Since the timebase's frequency stability is more important than a precise frequency, the values of the frequency determining components (C3, C4, R9, R10, and R11) are not terribly important.

For some applications (such as counting people passing through a doorway) the timebase oscillator can be eliminated. Also feel free to change the timebase frequency to suit your individual application.

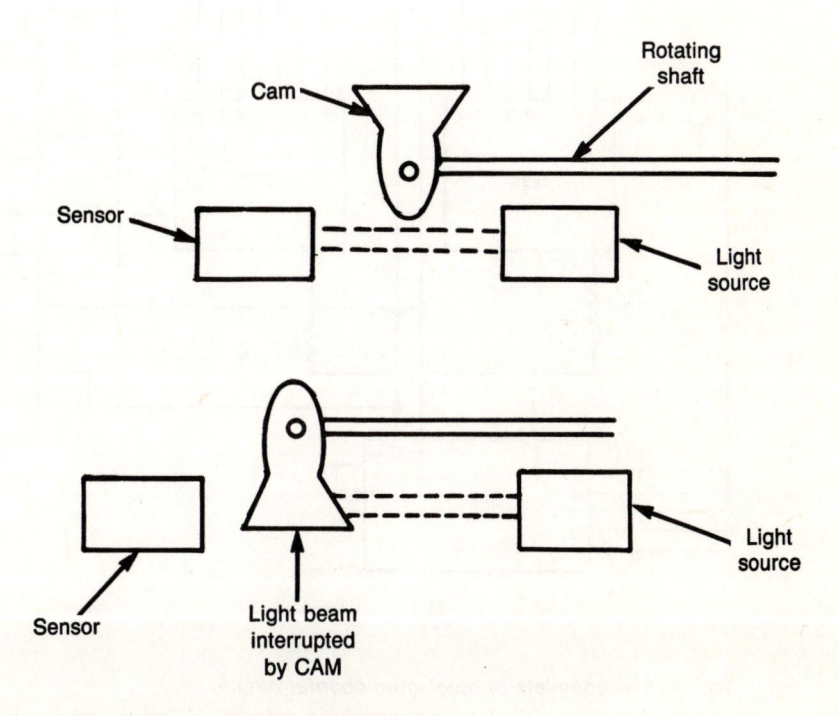

Fig. 7-19. A photoelectric rotation counter system.

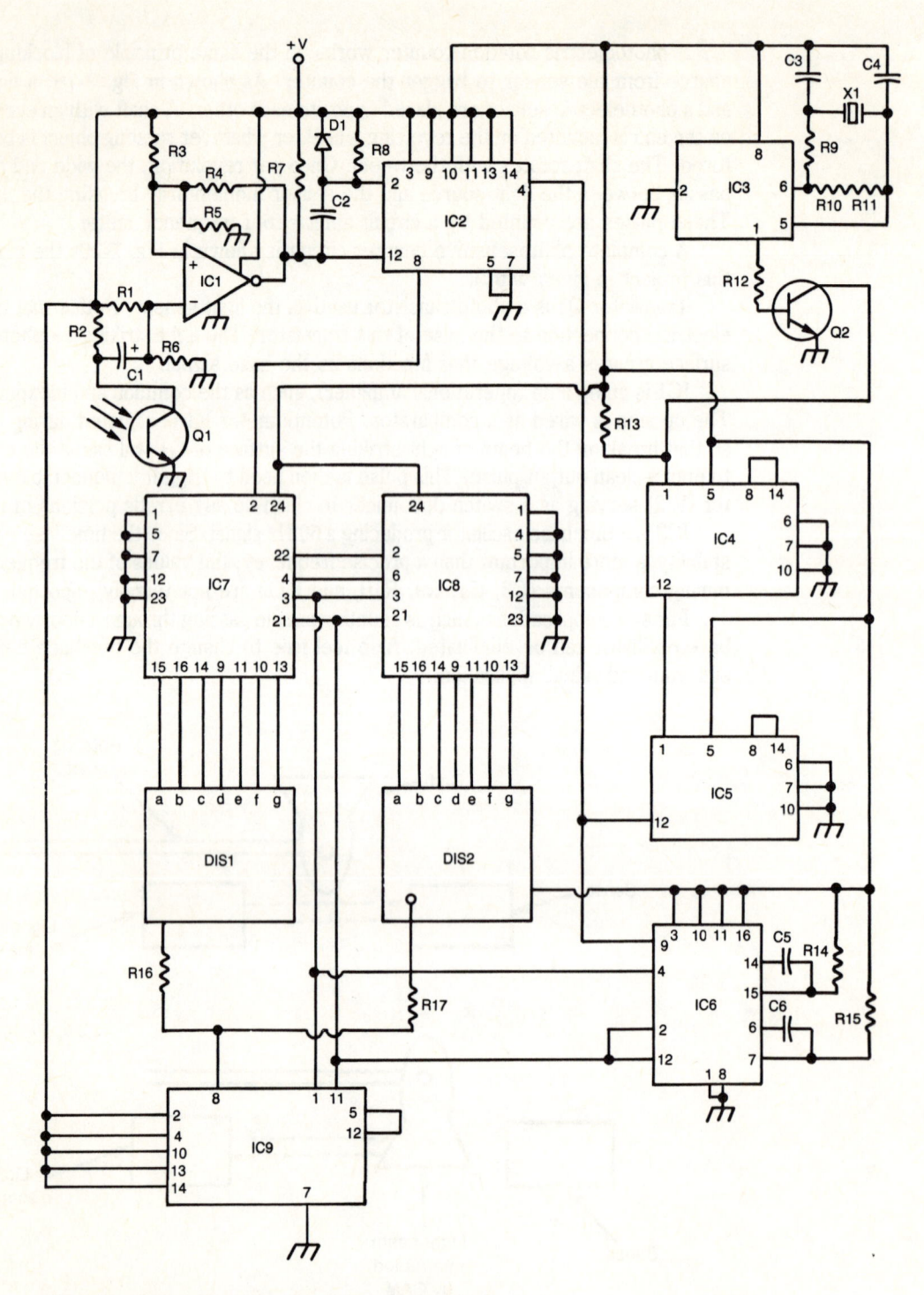

Fig. 7-20. A complete photoelectric counter circuit.

#35 PROJECT PARTS LIST

COMPONENT	DESCRIPTION
IC1	op amp (741, or similar)
IC2	74C90 J-K flip-flop
IC3	MM5369 60 Hz timebase
IC4, IC5	74C92 divide-by-12 counter
IC6	74C123 dual monostable multivibrator
IC7, IC8	74C143 decade counter/decoder/display driver
IC9	74C74 dual D flip-flop
DIS1, DIS2	seven-segment display, common anode (with decimal point)
D1	1N4148 diode (or similar)
Q1	FPT-100 phototransistor
Q2	NPN transistor (2N2222, 2N3904, or similar)
C1	1 µF capacitor
C2	1000 pF capacitor
C3	30 pF capacitor*
C4	6.2 pF capacitor *
C5, C6	0.033 µF capacitor
R1, R3, R5, R6	220kΩ resistor
R2	5.6kΩ resistor
R4	2.5 MΩ trimpot
R7, R9	1kΩ resistor
R8, R14, R15	10kΩ resistor
R10, R11	10 MΩ resistor *
R12	15kΩ resistor
R13	2.2kΩ resistor
R16, R17	330Ω resistor
X1	3.58 crystal (colorburst)

*see text

The actual counting is performed by IC4 and IC5. The display includes a decimal point, controlled by IC6 and IC9. Notice that the seven-segment LED display units used in this project are the common anode type. Common-cathode displays will not work with the 74C143 decoder/drivers (IC7 and IC8). Transistor Q2 can be almost any "garden-variety" NPN type device, such as the 2N2222, or the 2N3904.

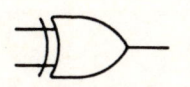

Project 36: Magnetic Reed Switch Counter

Magnetic reed switches are good "real-world" inputs for digital counters, and similar devices.

A magnetic reed switch is a small enclosed switch that is activated only when the presence of a sufficient magnetic field exists. Magnetic reed switches are available in both Normally Open, and Normally Closed configurations. For simplicity, we will only discuss Normally Open switches here. The project could be used with a Normally Closed switch, but the response to the magnetic field is reversed.

A switch is a switch, and it is easy enough to count the number of times the reed switch is closed. (A switch debouncing stage will probably be necessary.) In effect, we can count how many times a magnet is brought near the reed switch unit.

Why would we want to count such an occurrence? There are actually a great many potential applications for this project. For example, we could mount a permanent magnet on a revolving wheel or disc, so that it passes close to the magnetic reed switch once per revolution. This is illustrated in Fig. 7-21.

The counter will keep track of the number of revolutions of the wheel. By adding a timebase oscillator, we can determine how many times the wheel revolves per time period. Combining this information with the circumference of the wheel, we can determine its spinning speed. This concept can be used to construct an electronic speedometer for a bicycle, go-cart, or even a car.

To calibrate this type of speedometer, we must know the precise circumference (distance around the outer edge) of the wheel. We can then calculate how many revolutions there will be within a given distance of travel.

Another possible application for a magnetic reed switch counter is to measure wind speed. A wind vane with air-catching cups can be set up to turn a shaft with a magnet mounted on it. The digital circuitry counts the number of shaft revolutions.

Calibrating a wind speed circuit can be a bit tricky because the circumference of the shaft is so small. Probably the most efficient method is to enlist the aid of a friend who can drive a car while you calibrate the circuit. Do not try to drive and perform the

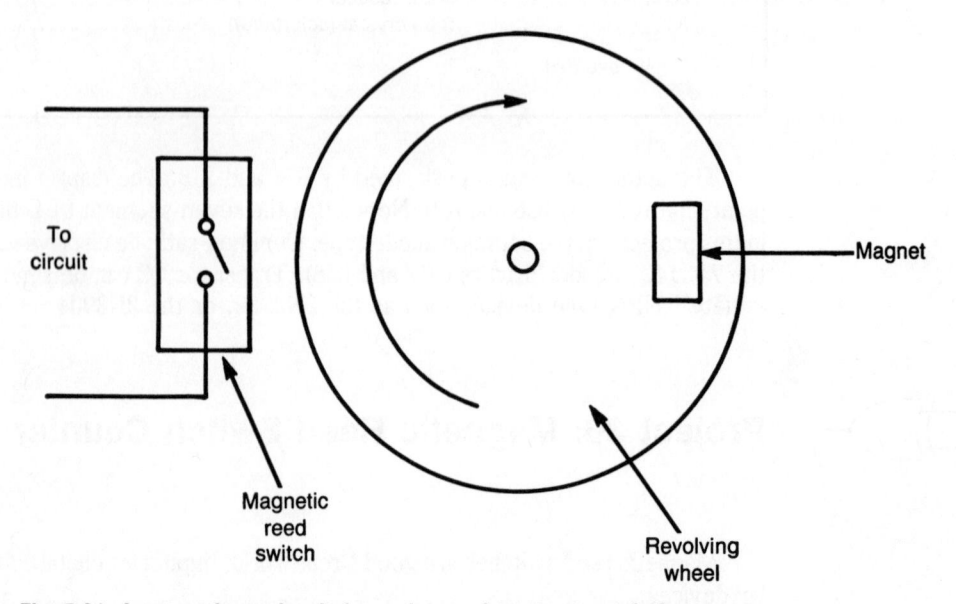

Fig. 7-21. A magnetic reed switch can be used to count revolutions.

calibration all by yourself. A calm day is needed for accurate calibration. Ideally, there should be no wind at all (0 mph windspeed). Unfortunately for our purposes here, such conditions are very rare. There is almost always some wind. If there is just a light breeze, you can still perform the calibration by driving at a 90° angle to the wind direction (or as close to 90° as you can get). For instance, if the wind is from the north, try to drive east or west. This way the existing breeze will not interfere noticeably with your calibration readings. If the prevailing winds are high they will throw off the calibration, no matter what direction you drive in.

While the driver is maintaining a steady speed, hold the wind vane out the window so it can rotate freely in the wind around the moving vehicle. Now simply adjust the circuit's calibration control(s) to get a reading that corresponds to the speed of the car.

This method is not exactly precise, there is a considerable margin for error from a number of sources. It should be close enough for most hobbyist applications. The difference between say, 11.6 mph and 12.3 mph winds is not likely to be very significant.

Magnetic reed switches are often used in burglar alarm systems to monitor whether a door or window is opened or closed. A permanent magnet is mounted on the door or window itself, while the switch is mounted on the frame, so the switch is activated when the door is closed, or deactivated when it is opened. We can monitor the traffic through a door by counting how many times it is opened and closed.

A block diagram for a practical magnetic reed switch counter circuit is shown in Fig. 7-22. Each time the magnetic reed switch is closed it triggers a monostable multivibrator functioning as a switch debouncer. This signal is gated with a timebase oscillator, and the counting takes place in the usual manner. If you are monitoring the traffic through a door, or using the circuit in some other application, you might want to eliminate the gate and timebase oscillator. In such applications, you just have to count the number of pulses at the output of the switch debouncer.

The actual schematic diagram for this project is shown in Fig. 7-23 with the parts list.

Potentiometer R24 is used to calibrate the circuit's readout. Closing switch S2 puts the counter into a test/reference mode, so you can easily determine when recalibration is required.

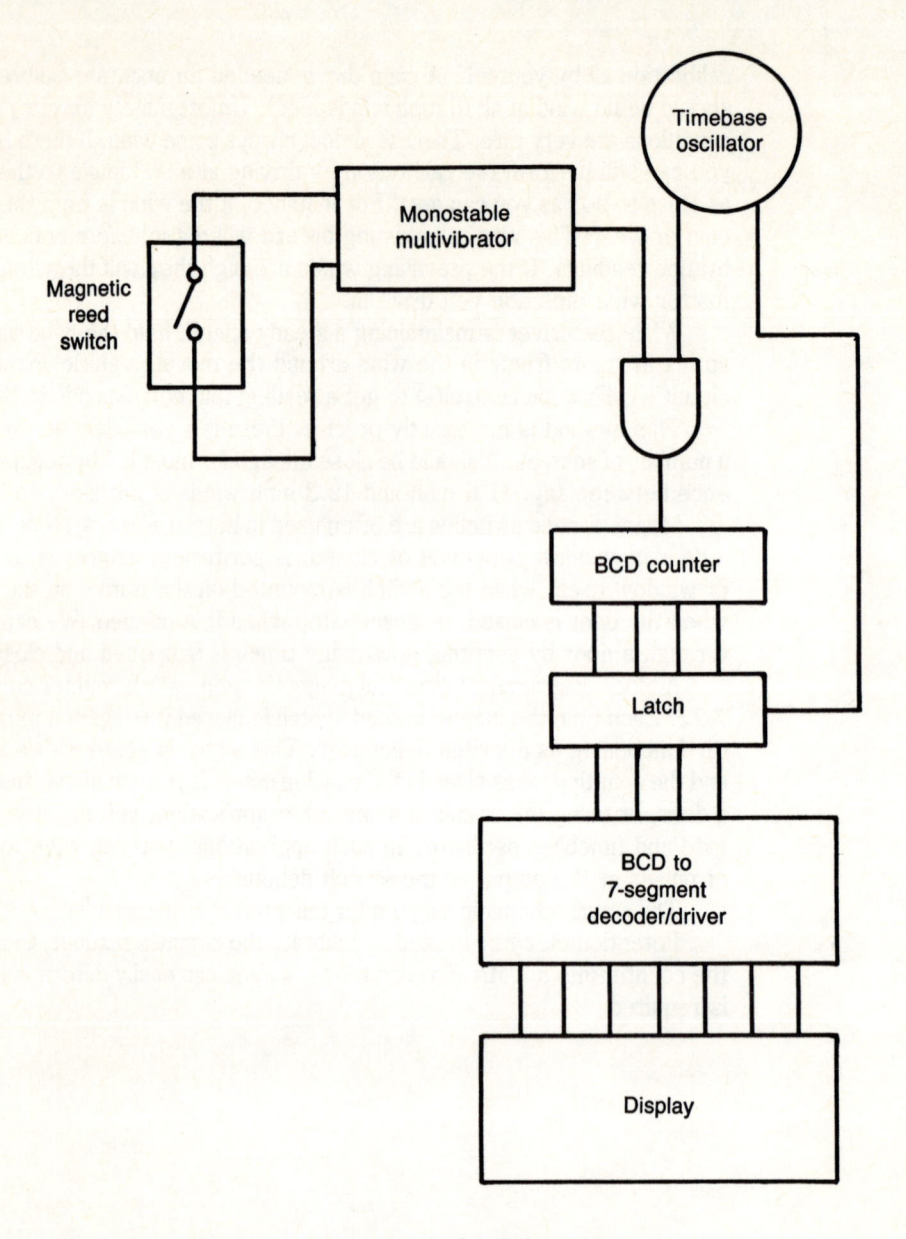

Fig. 7-22. A block diagram for a magnetic reed switch counter.

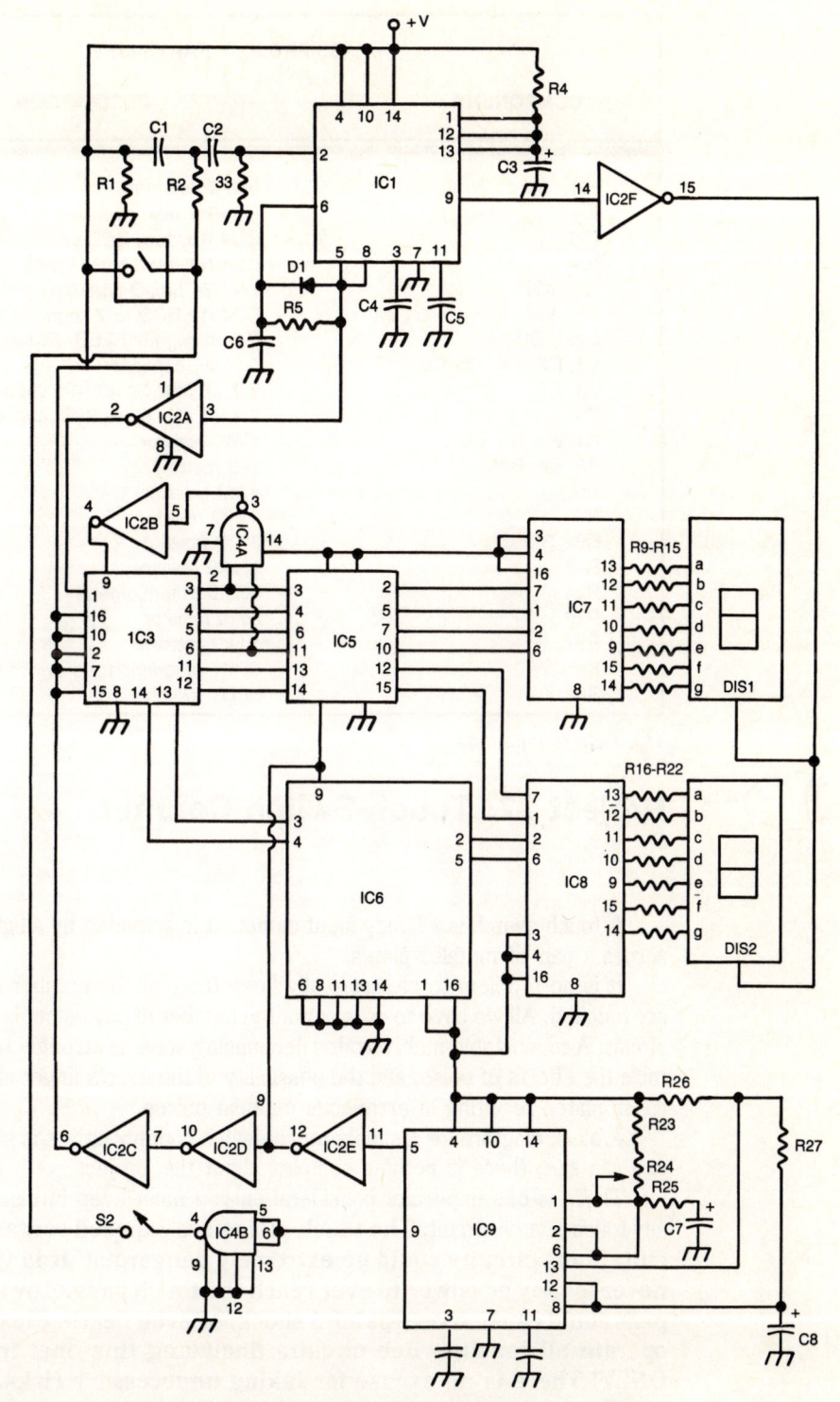

Fig. 7-23. The schematic for the magnetic reed switch counter circuit.

#36 PROJECT PARTS LIST

COMPONENT	DESCRIPTION
IC1, IC9	556 dual timer
IC2	CD4049 hex inverter
IC3	CD4518 dual BCD counter
IC4	CD4011 quad NAND gate
IC5, IC6	74C174 hex D-type flip-flop
IC7, IC8	CD4511 BCD to 7-segment decoder/driver
DIS1, DIS2	Seven-segment LED display, common cathode
C1, C2, C4, C5, C6	0.01 µF capacitor
C3, C7	4.7 µF 35V electrolytic capacitor
C8	2.2 µF 35V electrolytic capacitor
R1, R3, R4, R5	1 MΩ resistor
R2, R8, R25	1kΩ resistor
R6	47kΩ resistor
R7	2.2kΩ resistor
R9 - R22	330Ω resistor
R23	100kΩ resistor
R24	500kΩ potentiometer
R26	56kΩ resistor
R27	1.5kΩ resistor
S1	SPST magnetic reed switch (see text)
S2	SPDT switch

Fig. 7-23. (continued)

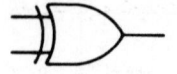

Project 37: Touch-Switch Counter

A touch-switch is a handy input circuit. It is activated by a light touch of a fingertip across a pair of metallic plates.

It is no problem to electronically keep track of the number of times the contacts are touched. All we have to do is count the number of pulses put out by the touch-switch circuit. A monostable multivibrator debouncing stage is strongly recommended to minimize the effects of noise, and the possibility of the user's finger shifting position on the touch plates resulting in erroneous multiple pulses.

A block diagram for a simple touch-switch counter circuit is shown in Fig. 7-24. As you can see, there is nothing complex about this project.

There is one important consideration you must keep in mind when working with any touch-switch circuit. **The touch plates are exposed contact points in the circuit. Short circuits could be extremely dangerous. It is vitally important to never let any ac power to ever reach the touch pads. Any ac across the touch pads could cause a very painful shock, or even death. Please play it safe, and operate all touch-switch circuits (including this one) from battery power ONLY! There is no excuse for taking unnecessary risks.**

The actual schematic diagram for this project, and its parts list appears in Fig. 7-25.

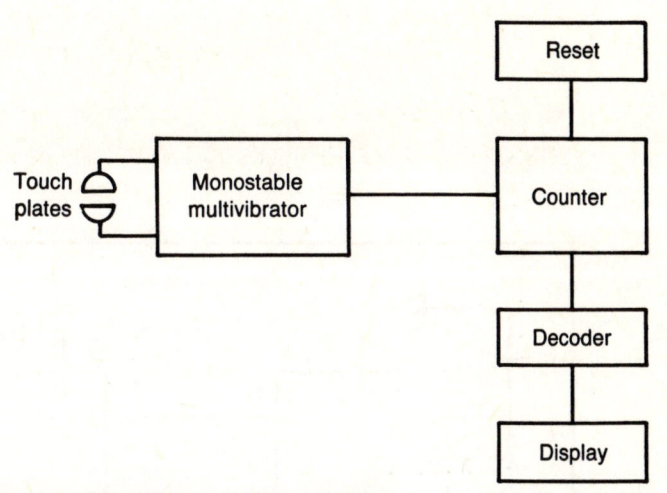

Fig. 7-24. A touch switch used to advance a counter.

When you touch your finger across the two touch pads, your body conducts enough electricity to introduce a signal into the circuit. Because the body serves as a conductor for the circuit, any possibility for a dangerous ac voltage across the touch pads must be absolutely eliminated. The input signal is inverted by IC1A, and fed into the trigger input of the monostable multivibrator stage (IC2).

The monostable multivibrator essentially *stretches* the input pulse, functioning as a switch debouncer. IC2 feeds a clean, reliable signal to the input of the counter (IC3).

The counter is a basic BCD two digit (00 to 99) decimal counter. Additional stages can be added for higher count values. The counter can be reset back to all zeroes by momentarily moving switch S1 from ground to +V.

The count will be incremented (increased by one) each time the touch pads are touched, unless the counter is manually reset via S1, or the maximum count (99) is exceeded.

It wouldn't be much trouble to adapt this circuit to include a timebase oscillator. With this modification, the circuit will determine how many times the touch switch is activated within a specific time period. The additional circuitry required for this modification is shown in Fig. 7-26, a supplementary parts list appears with it.

To make this modification, make a break in the original circuit (Fig. 7-25) at the point marked "X." The output signal from IC2 pin #3 is now fed to pin #12 of IC1D. This NANDs the input signal with the output of the timebase oscillator (IC6). An inverter stage (IC7A) converts the NAND operation into an AND function. Now an output pulse signal is fed to the counter input (IC3, pin #1) *only* when the timebase oscillator is at logic 1.

The timebase oscillator also takes care of automatically resetting the counter at the end of each cycle, replacing switch S1 at IC3 pins #7 and 15. The two inverter stages in this line (IC7B and IC7C) are included to slow down the reset signal slightly, allowing time for the count to be displayed before the counter is reset to 00.

Depending on your specific application, you might want to keep the manual reset switch (S1) in parallel with the automated reset signal. To be functional, the timebase oscillator should have a very, very low frequency.

146

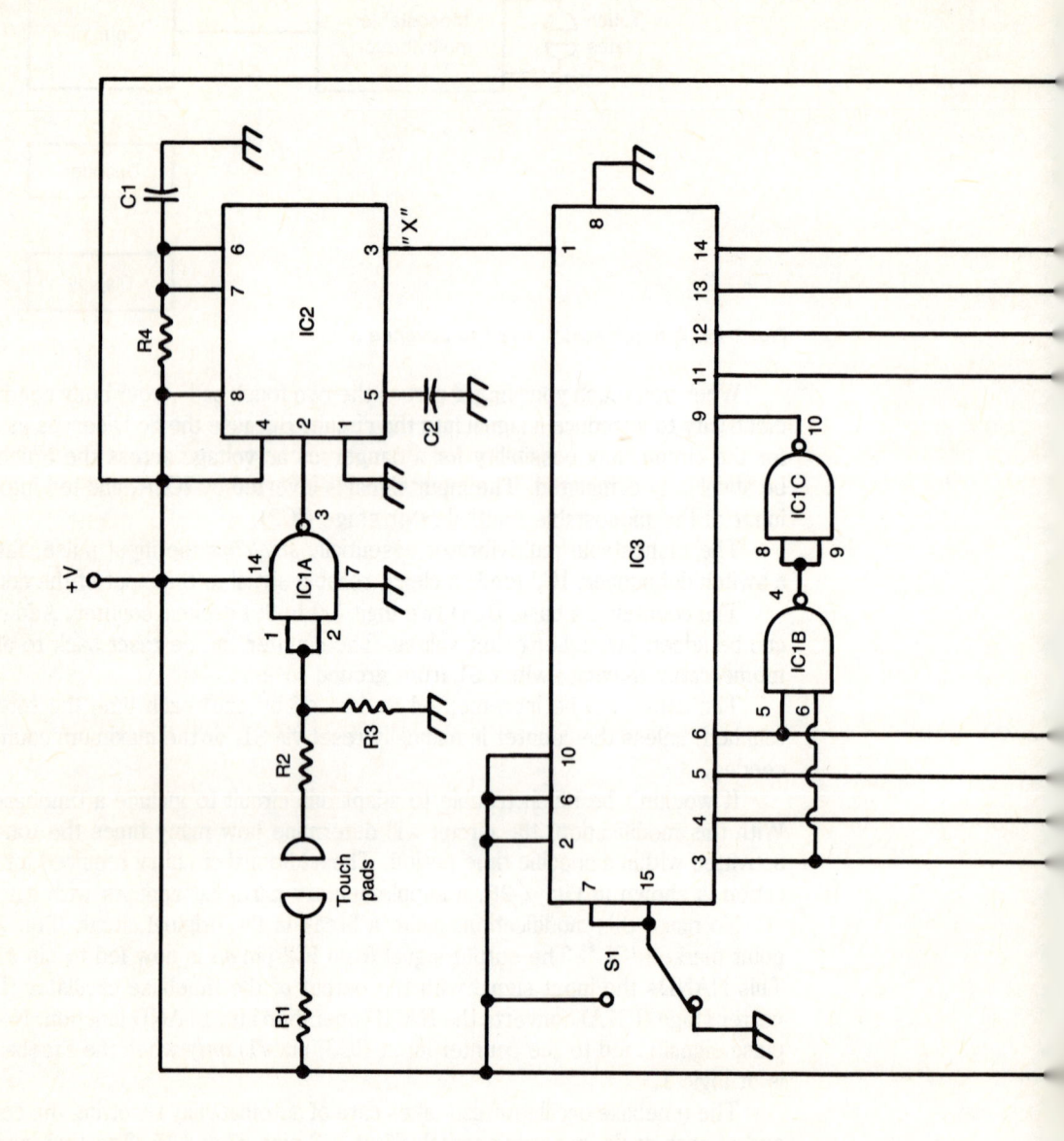

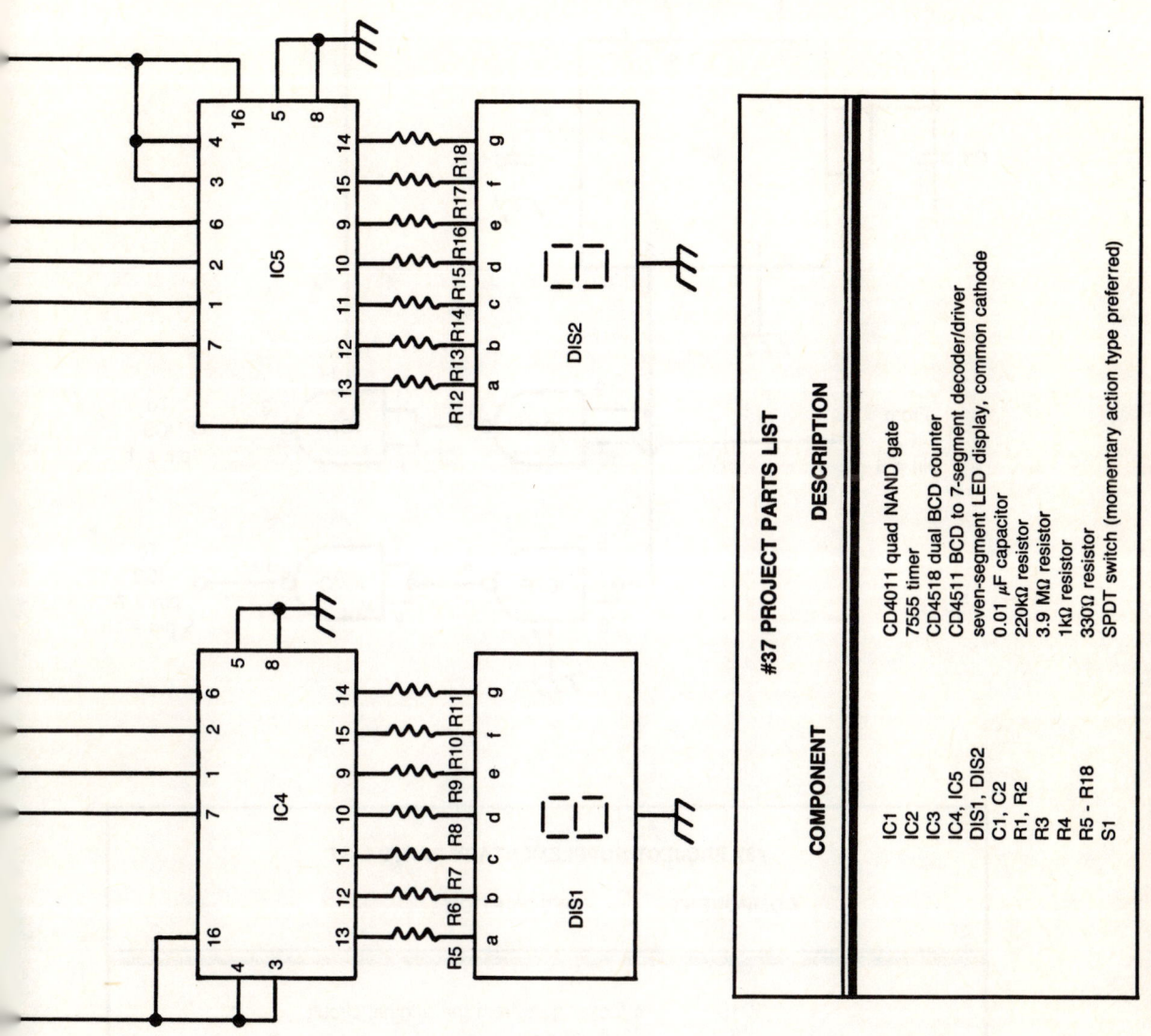

Fig. 7-25. The complete schematic for a touch switch counter circuit.

#37 PROJECT PARTS LIST

COMPONENT	DESCRIPTION
IC1	CD4011 quad NAND gate
IC2	7555 timer
IC3	CD4518 dual BCD counter
IC4, IC5	CD4511 BCD to 7-segment decoder/driver
DIS1, DIS2	seven-segment LED display, common cathode
C1, C2	0.01 µF capacitor
R1, R2	220kΩ resistor
R3	3.9 MΩ resistor
R4	1kΩ resistor
R5 - R18	330Ω resistor
S1	SPDT switch (momentary action type preferred)

147

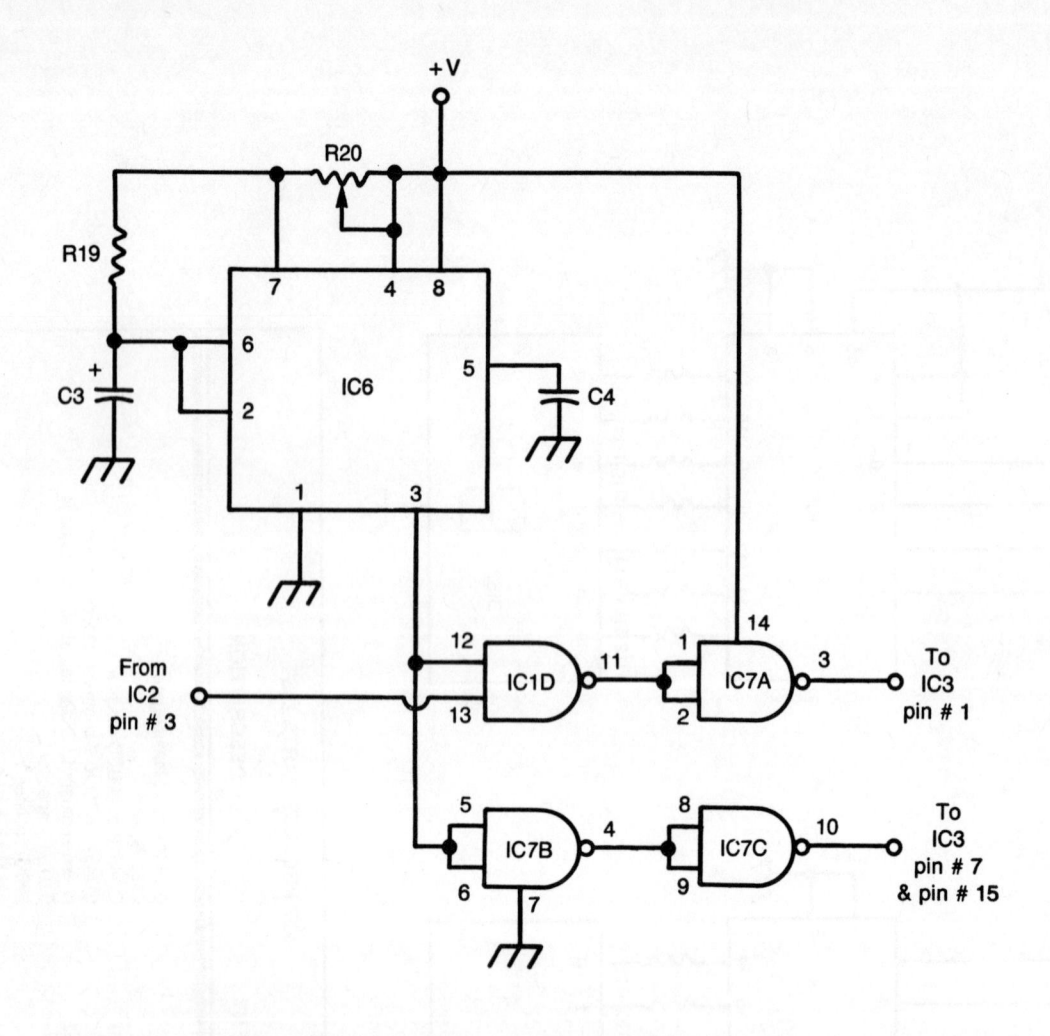

#37 PROJECT SUPPLEMENTARY PARTS LIST

COMPONENT	DESCRIPTION
IC1D	left-over gate from the original circuit
IC6	7555 timer
IC7	CD4011 quad NAND gate
C3	100 µF 35V electrolytic capacitor
C4	0.01 µF capacitor
R19	10kΩ resistor
R20	500kΩ potentiometer

Fig. 7-26. This modification adds a timebase to the circuit of Fig. 7-25.

8

TIME KEEPING CIRCUITS

A counter circuit, like most of those presented in Chapter 7, can keep track of how many pulses have appeared at its input. If we use a continuous series of regular pulses at a fixed frequency, the number of pulses can be used to determine the amount of time that has passed. This simple trick is put to work in the time keeping circuits of this chapter.

Counter circuits can be used to build digital clocks and timers. This time we mean "clock" in the usual sense—a time-keeping device.

The first thing you need when designing an electronic timepiece is an accurate timebase. A timebase is basically the same as the clock oscillators used to keep various digital circuits in synchronization within a system. A digital time-telling clock essentially counts the number of seconds, minutes, and hours. To do this, it obviously needs some way to know how long a second is. This is the job of a timebase. The timebase generates a signal with a very precise frequency so that the circuitry can reliably count X number of pulses per second.

This timebase frequency must be <u>extremely</u> accurate. Even a slight error can destroy the usefulness of the system. It might seem that there is little difference between 0.95 seconds and 1.00 second. But remember that in time-keeping applications the error is cumulative. Each and every second is 5 percent short. When you multiply that 5 percent error over a 24-hour day you wind up with a day that lasts only 22 hours and 48 minutes. That's clearly not very good time-keeping. About the only thing that clock would be good for is paperweight duty.

Most electronic clock circuits work with a timebase of 60 Hz (60 pulses per second). This is really more due to tradition and habit, than any technological reason, especially in battery operated equipment. This tradition stems from the ac electric power lines that operate at a frequency of 60 Hz (in the U.S.). In many early electric clocks the ac line frequency was used as a convenient source of a timebase signal. This was fine for electromechanical clock motors, but it is not suitable for digital circuits. Adapting the ac power signal for use in a digital circuit is difficult at best. With modern technological alternatives, it is certainly more trouble than it's worth to even try.

Most digital clocks work with some sort of crystal oscillators. Quartz crystals can hold to a very precise frequency.

Incidentally, do you remember when digital watches first came out? Many models (especially the more expensive ones) were proudly touted in the ads as being quartz controlled, as if this was a special feature. *All* digital watches are quartz controlled. A crystal oscillator is used for the timebase, and a crystal is a slab of quartz. I don't know of any other practical or economical way to do the job. The ads were obviously playing of the public's ignorance of what the term "quartz controlled" meant. But it certainly sounded impressive.

Most crystals have resonant frequencies much higher than the desired 60 Hz. Generally, crystal resonant frequencies are above 1 MHz (1,000,000 Hz), which is far too high for direct use in a digital clock.

The solution is to add one or more counter stages to drop the crystal's frequency down to the desired value. You could use separate counter ICs for each necessary stage, but you will end up with an unnecessarily bulky, and expensive circuit. Fortunately there is an easier way. As is often the case for popular digital functions, there is a dedicated IC available to do the job. The MM5369 60 Hz timebase chip is illustrated in Fig. 8-1. This device generates an extremely precise, stable 60 Hz timebase signal from a 3.579545 MHz crystal. Why this seemingly oddball frequency? This frequency was selected because it is one of the most readily available crystal frequencies. This frequency (which is often rounded off to 3.58 MHz for convenience) is used in color TV sets to provide the color burst signal. 3.58 MHz crystals are easy to find, and are available from virtually every dealer who handles any crystals at all.

The MM5369 is a single eight-pin chip that provides all the necessary division to reduce a 3.58 MHz input signal to a 60 Hz output signal. Actually, except for the V_{DD}, and ground connections, only three of the IC's pins are used. Two of the connections

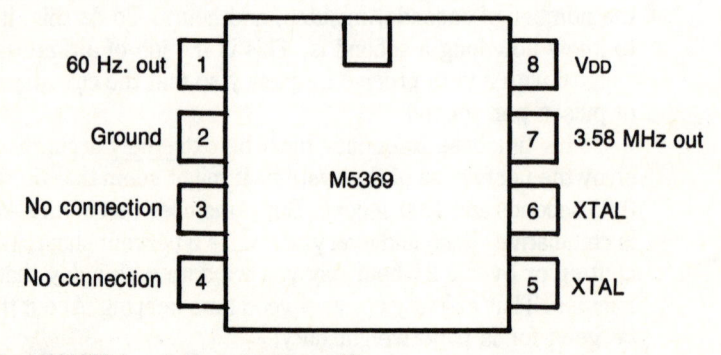

Fig. 8-1. The MM5369 is a 1 Hertz timebase IC.

are made to either end of the 3.58 MHz crystal. The third active lead is the output line for the 60 Hz timebase signal. Clearly, circuits build around the MM5369 are not particularly complex.

There is one additional active pin on this chip, which may or may not be used, depending on the specific application. Pin #7 provides a buffered 3.58 MHz output signal, just in case it might be needed in certain circuits.

The remaining two pins on the MM5369 (pins #3 and 4) are included just to make the IC fit the standard eight-pin DIP format. These two pins are not internally connected to any of the chip's circuitry.

A 60 Hz timebase signal is made up of sixty pulses per second. To increment the seconds counter once every second, we need exactly one pulse per second. To create a one second timer, we therefore must divide the 60 Hz timebase signal by sixty.

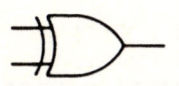

Project 38: One Hertz Timebase

Figure 8-2 shows a practical circuit for generating a precise 1 Hz output signal. A parts list for this project is given in the figure.

Notice that three resistors and two capacitors are included in a network around the crystal. These components are not absolutely essential, but they can greatly improve the stability of the signal. According to the manufacturer, the following capacitance values are preferred:

C_1 6.36 pF
C_2 30   pF

Unfortunately, neither of these capacitance values are commonly available. You can round the capacitance values off to standard values without significantly hurting the precision of the output frequency:

C_1 10 pF
C_2 47 pF

The error introduced by this rounding off will be slight enough to be ignored in most applications. If your specific application demands absolute precision, trimmer (variable) capacitors can be included in the circuit for fine tuning.

The 60 Hz timebase signal from the output of IC1 is fed to IC2, a CD4017 decade counter wired to divide by six, this drops the frequency to 10 Hz. IC3 is another CD4017 decade counter wired to divide by ten, resulting in a precise 1 Hz output signal.

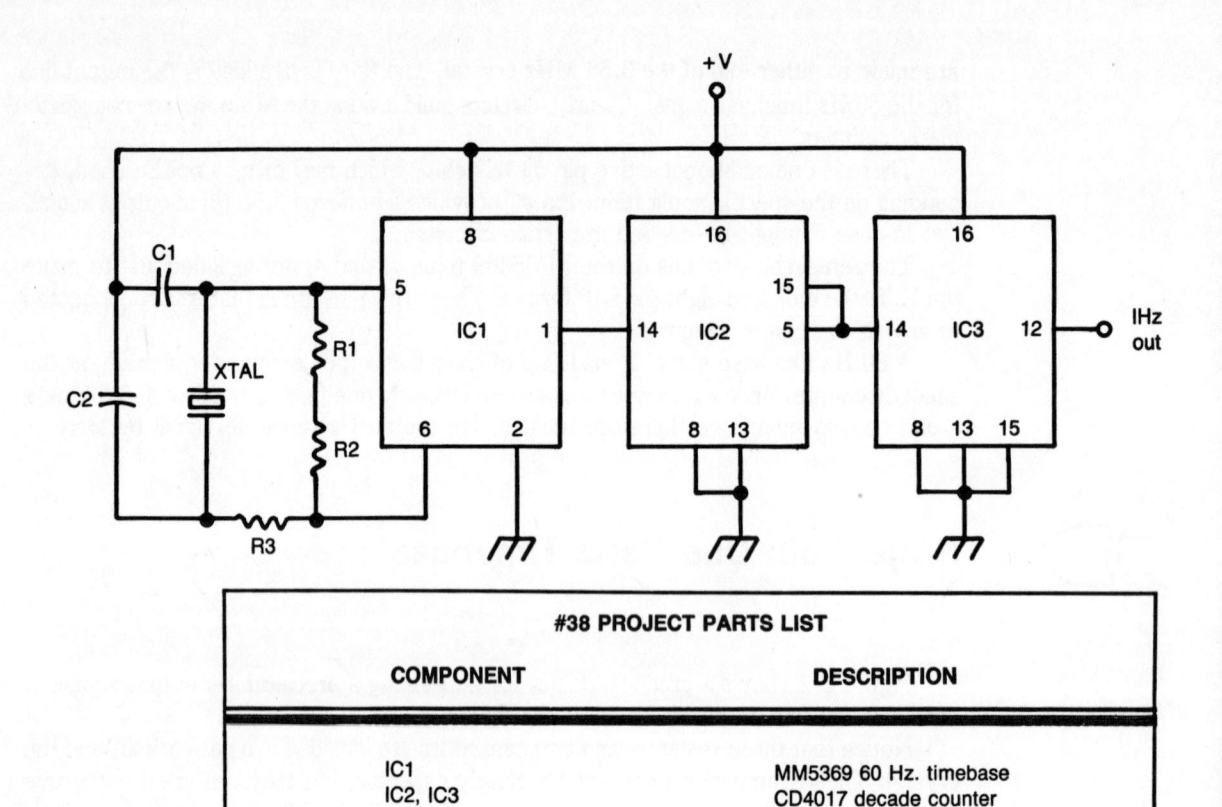

Fig. 8-2. A practical 1 Hz timebase circuit.

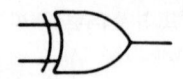

| #38 PROJECT PARTS LIST ||
COMPONENT	DESCRIPTION
IC1	MM5369 60 Hz. timebase
IC2, IC3	CD4017 decade counter
XTAL	3.58 MHz colorburst crystal
C1	*
C2	*
R1, R2	10MΩ resistor
R3	1kΩ resistor
* see text	

Project 39: One Minute Timer

Adding a handful of additional components to a one Hertz timebase (Project #38) allows us to create a practical timer circuit. The primary components are a pair of additional timers and an AND gate. Figure 8-3 shows a circuit for a 60 second (1 minute) timer. Notice that the 1 Hz timebase circuit presented earlier is included in this circuit too.

This project is very easy to use, when switch S1 is moved to the RUN position, sixty seconds will be counted, then the LED will turn on. The LED will remain lit until S1 is returned to the RESET position.

ICs 1 through 3 (and their associated components) make up a 1 Hz timebase. ICs 4 and 5 are two additional decade counters, to divide the timebase signal by sixty. Since the timebase (output of IC3) produces exactly one pulse per second, counting sixty pulses

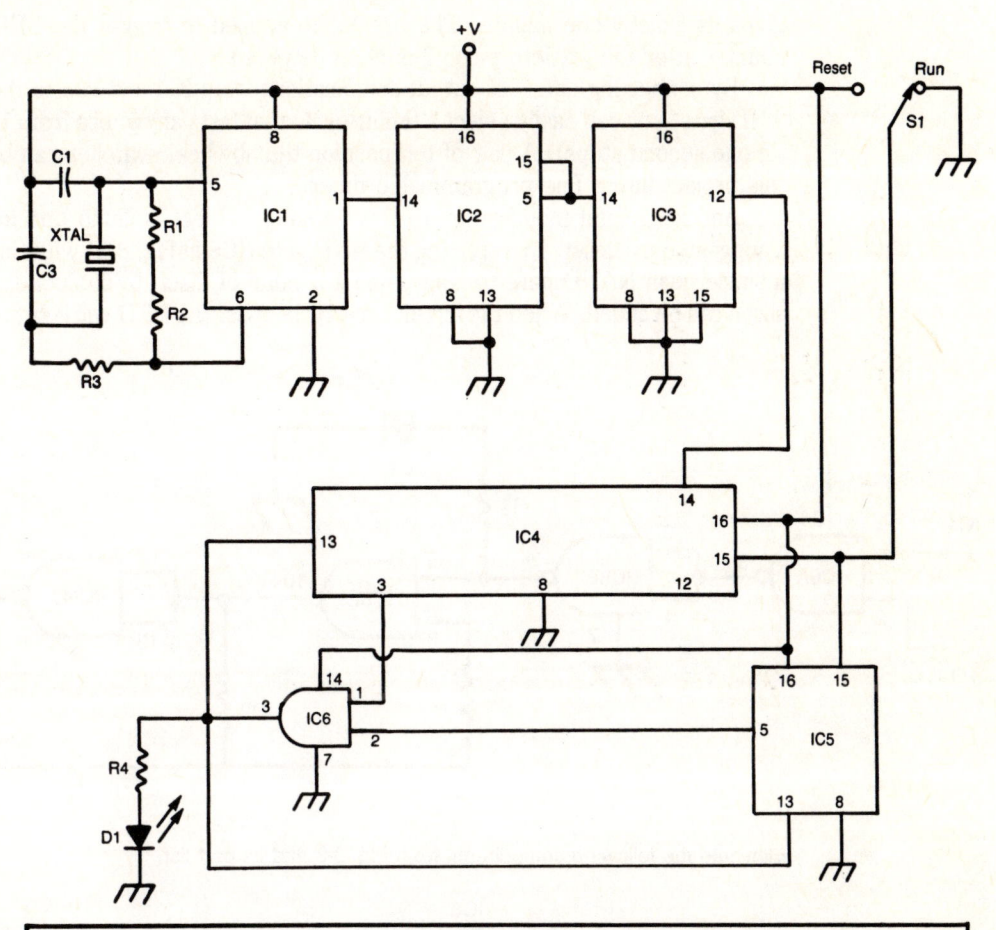

#39 PROJECT PARTS LIST

COMPONENT	DESCRIPTION
IC1	MM5369 60 Hz timebase
IC2 - IC5 CD4017 decade counter	
IC6*	CD4081 quad AND gate
D1*	LED
XTAL	3.58 MHz
C1	10 pF capacitor
C2	47 pF capacitor
R1, R2	10MΩ resistor
R3	1kΩ resistor
R4*	330 Ω resistor
S1	SPDT switch

* these components are to be eliminated for the audible alarm version shown in Fig. 8-4.

Fig. 8-3. A one minute timer.

gives us exactly one minute. The AND gate is used to trigger the LED and stop the counter after the sixtieth pulse has been detected.

By connecting different output combinations from IC4 and IC5 to the inputs of the AND gate, you can easily select a timing cycle that lasts anywhere from 1 to 99 seconds (in one second steps). A pair of ten position thumbwheel switches can be used to turn this project into a fine programmable timer.

Another useful modification to this versatile project is illustrated in Fig. 8-4 with a suggested parts list. By replacing the AND gate (IC6 in Fig. 8-3) with this circuit (using a single quad NAND gate package, and a handful of discrete components), an audible alarm can be added. When the circuit time cycle ends, the LED will light, as in the earlier

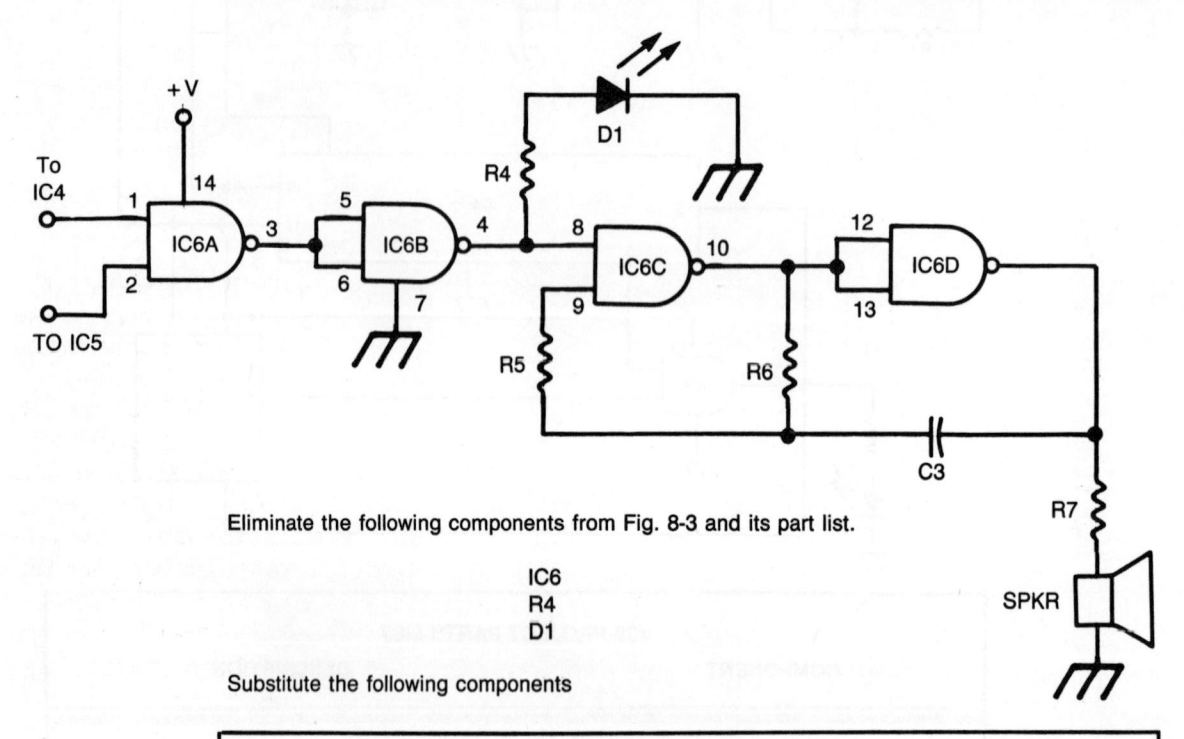

Eliminate the following components from Fig. 8-3 and its part list.

IC6
R4
D1

Substitute the following components

#39 PROJECT PARTS LIST

COMPONENT	DESCRIPTION
IC6	CD4011 quad NAND gate
D1	LED
C3	0.01 µF disc capacitor
R4	330Ω resistor
R5	1MΩ resistor
R6	100kΩ resistor
R7	100Ω resistor
SPKR	small speaker

Fig. 8-4. An audible alarm added to the timer circuit of Fig. 8-3.

version, and a tone will be sounded. The LED will remain lit, and the tone will continue to sound until switch S1 is returned to the RESET position.

IC6A and IC6B work together to duplicate the action of the original AND gate. IC6C and IC6D are wired as a gated tone generator with an output of approximately 1 kHz (1000 Hz). The tone generator can oscillate only when a HIGH logic signal is applied to pin #8.

The value of R7 can be adjusted to control the volume of the tone emitted by the speaker. The smaller this resistor is, the louder the tone will sound. To prevent damage to the speaker, it is probably best not to make R7 lower than 50 ohms. At the other extreme, if resistor R7 is made larger than about 1000 ohms, the tone will probably be too weak to be audible.

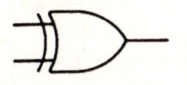

Project 40: Digital Clock

The 1 Hz timebase project (#38) can also serve as the heart of the complete digital clock. It's just a matter of adding the appropriate counters and displays. For convenience, the circuit is broken down into three sections. The first section is the 1 Hz timebase, and is exactly the same circuit we used for Project #38.

The second section of the digital clock project is shown in Fig. 8-5. IC4 and IC5 divide the one Hertz pulses by sixty. The output of IC5 produces a pulse once every minute. IC6, a dual BCD counter, tallies the number of minutes that have passed. The circuit will display minute values from 00 to 59. Once the count reaches 60, the counters are reset to 00. The count is displayed on a pair of seven-segment LED displays.

Figure 8-6 shows the circuitry for adding the display for hours, ranging from 01 to 12. The ones digit is wired in the usual fashion, but we can take a few short cuts for the tens column. This digit will always be either a 0 or a 1. Segments b and c will always be lit, so they are tired directly to the positive power supply line (through current dropping resistors, of course).

If the hours count is less than 10, then segments a, d, e, and f are lit to produce a 0 on the display. Otherwise, a 1 will be displayed. Segment g is never lit.

Note that the hours counter must be reset to 01, not 00.

The complete block diagram for this digital clock project is shown in Fig. 8-7, including the parts list. Notice that the components for the 1 Hz timebase are omitted from the parts list. Project #38 is used in its entirety.

This is a very simple digital clock circuit with no special features. Switches S1 and S2 are used to manually set the time. Pressing one of these switches will advance the counter one step per second, as long as the switch is held closed. Be aware that setting the minutes past 59 will advance the hours counter. It is best to set the minutes first, then set the hours.

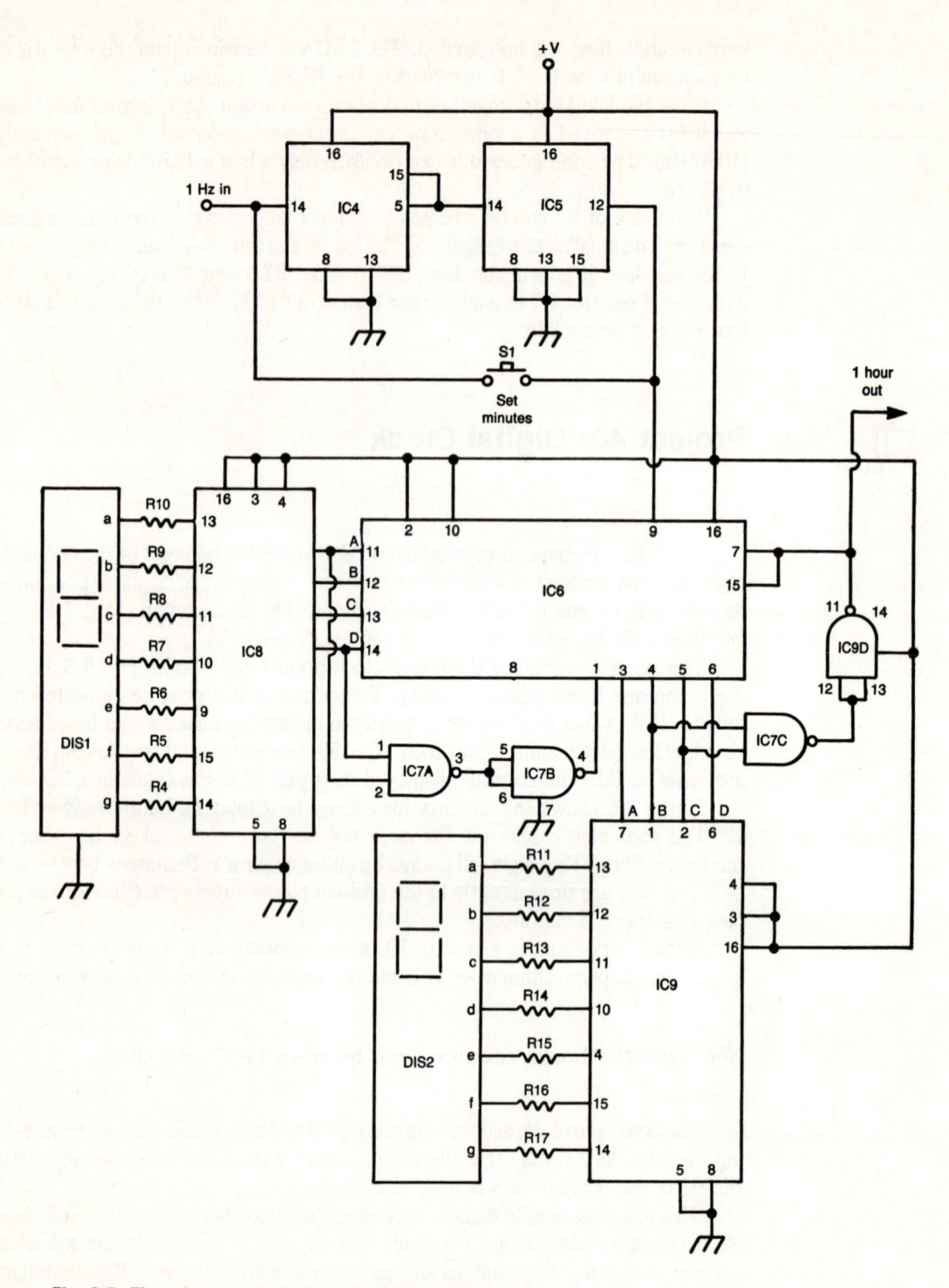

Fig. 8-5. The minutes section of the digital clock project.

#40 PROJECT PARTS LIST

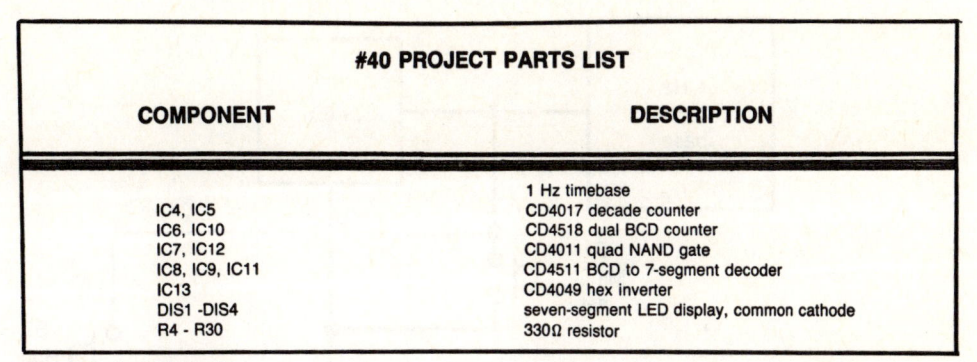

COMPONENT	DESCRIPTION
	1 Hz timebase
IC4, IC5	CD4017 decade counter
IC6, IC10	CD4518 dual BCD counter
IC7, IC12	CD4011 quad NAND gate
IC8, IC9, IC11	CD4511 BCD to 7-segment decoder
IC13	CD4049 hex inverter
DIS1 -DIS4	seven-segment LED display, common cathode
R4 - R30	330Ω resistor

Fig. 8-5. (continued)

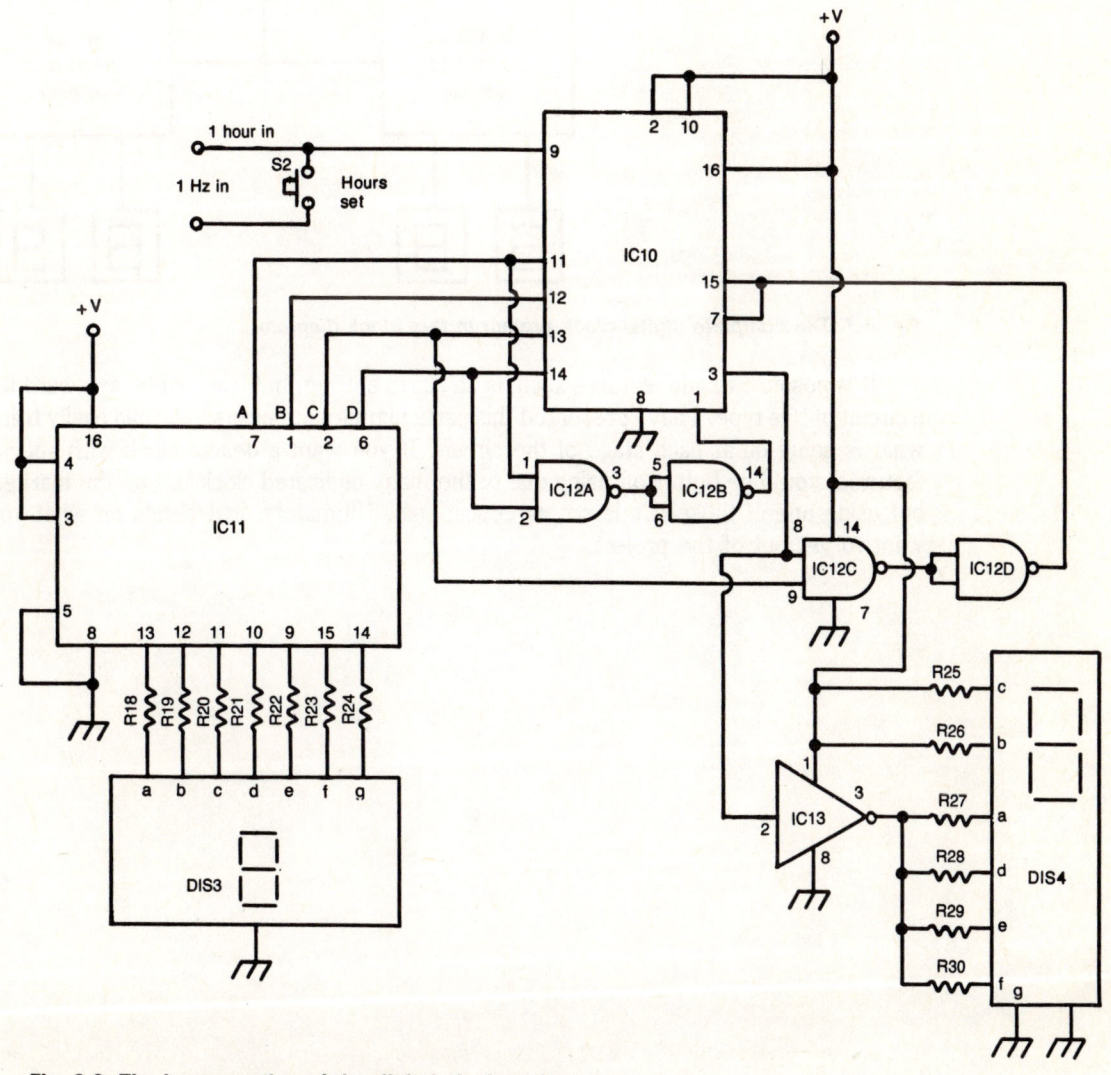

Fig. 8-6. The hours section of the digital clock project.

157

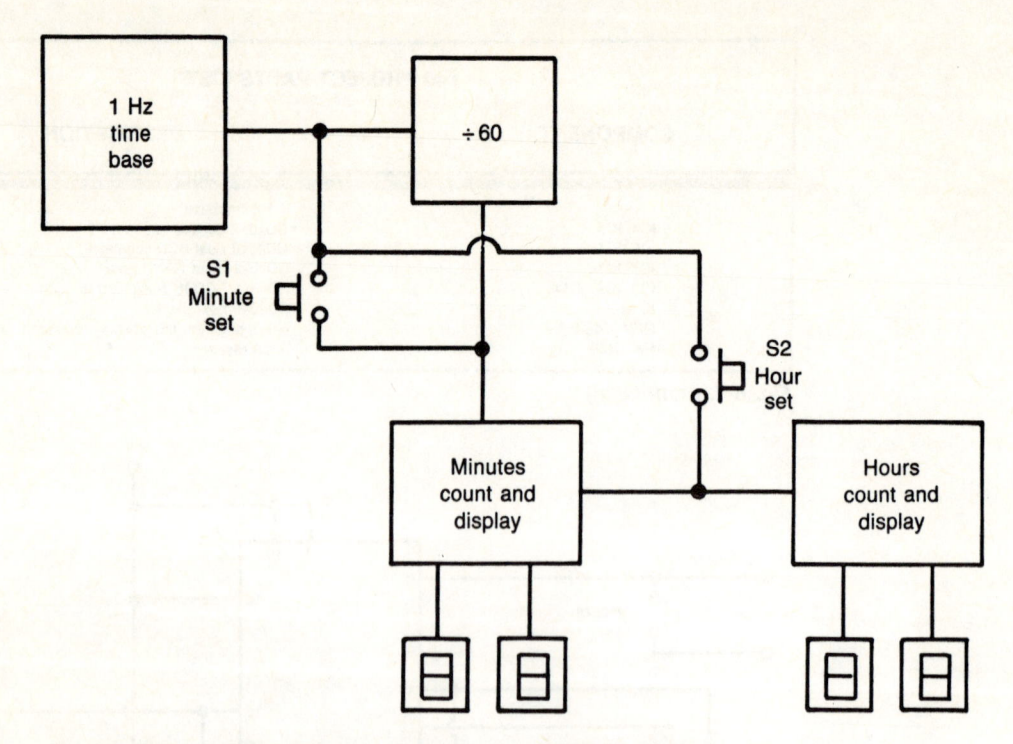

Fig. 8-7. The complete digital clock project in this block diagram.

It is possible to add features such as an alarm setting, but it would be awkward for a circuit of this type. I have presented this particular circuit because you can easily trace what is going on at each stage of the circuit. If you want a deluxe clock with special features, you'd be better off using one of the many dedicated clock ICs on the market, but using one of those ICs is not as educational. Ultimately, it depends on what you want to get out of the project.

9
GAME CIRCUITS

If you are only interested in serious applications, by all means skip this chapter, this one is just fun. This chapter features a pair of complete electronic game circuits, I certainly hope you'll have some fun with them.

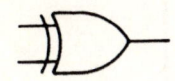

Project 41: Electronic Dice

An earlier project was a random number generator, we can extend and modify this idea slightly to create a set of electronic dice. Besides being fascinating, and entertaining on its own, this project can be used in many games where the play is determined by a roll of the dice.

The big difference from the basic random number generator discussed earlier is that the count for each die ranges from 1 to 6. Individual LEDs are used to represent the spots of the die. These LEDs are arranged in the traditional dice pattern, as shown in Fig. 9-1. Notice that seven LEDs are required, although no valid count will light them all at the same time.

The circuitry for a single die is shown in Fig. 9-2, accompanied by the parts list. For most applications you will probably want to build a pair of dice circuits. They can

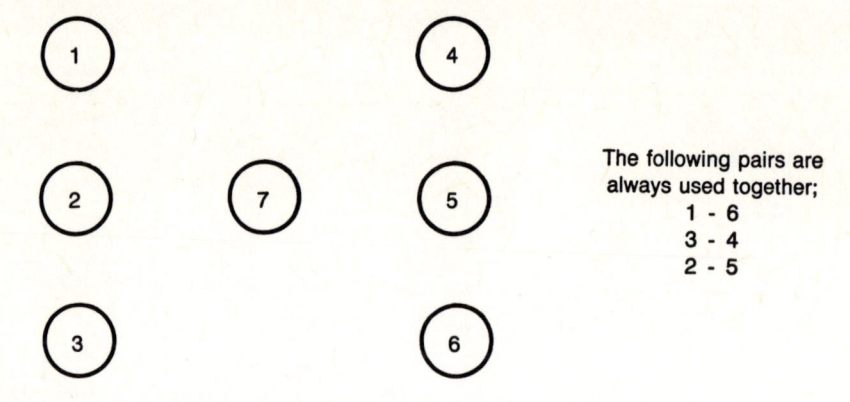

```
    (1)                    (4)

                                        The following pairs are
    (2)        (7)        (5)           always used together;
                                              1 - 6
                                              3 - 4
                                              2 - 5
    (3)                    (6)
```

Fig. 9-1. Seven LEDs used to simulate the face of a die.

be built into a single housing, but do not try to economize by driving both dice with a single clock, or the dice will always be the same.

In this circuit a CD4017 decade counter chip is connected as a modulo-six cycling counter. The gates of IC3 and IC4 determine which LEDs will be lit for each of the six possible count values.

While the clock is feeding the counter (i.e., switch S1 is closed), all seven output LEDs will appear to be continuously lit. If the LEDs are positioned as shown in the figure, the counter will always stop on a valid die face pattern. Be very careful to follow the numbering exactly, or you might get some odd output patterns. The count (number of dots lit) will be correct, but the display will not resemble a die, eliminating much of the point of the project.

For each count the LEDs should be lit in the patterns given in Table 9-1:

COUNTER OUTPUT	LEDs 1 2 3 4 5 6 7
1	– – – – – – X
2	X – – – – X –
3	X – – – – X X
4	X – X X – X –
5	X – X X – X X
6 (0)	X X X X X X –

Table 9-1.

where "X" represents a lit LED, and "–" represents an LED that remains dark. The patterns are illustrated in Fig. 9-3.

Many people find roulette an exciting game of chance. Like the electronic dice circuit, this project is built around a random number generator.

This electronic version works just like a regular, mechanical roulette wheel, except the winning spot is determined by a lit LED, rather than a ball falling into a hole. When the clock stops, only one LED will be lit. All of the other LEDs will be dark. For the most realistic action, the LEDs should be arranged in a ring to simulate a roulette wheel.

Fig. 9-2. The complete electronic dice circuit.

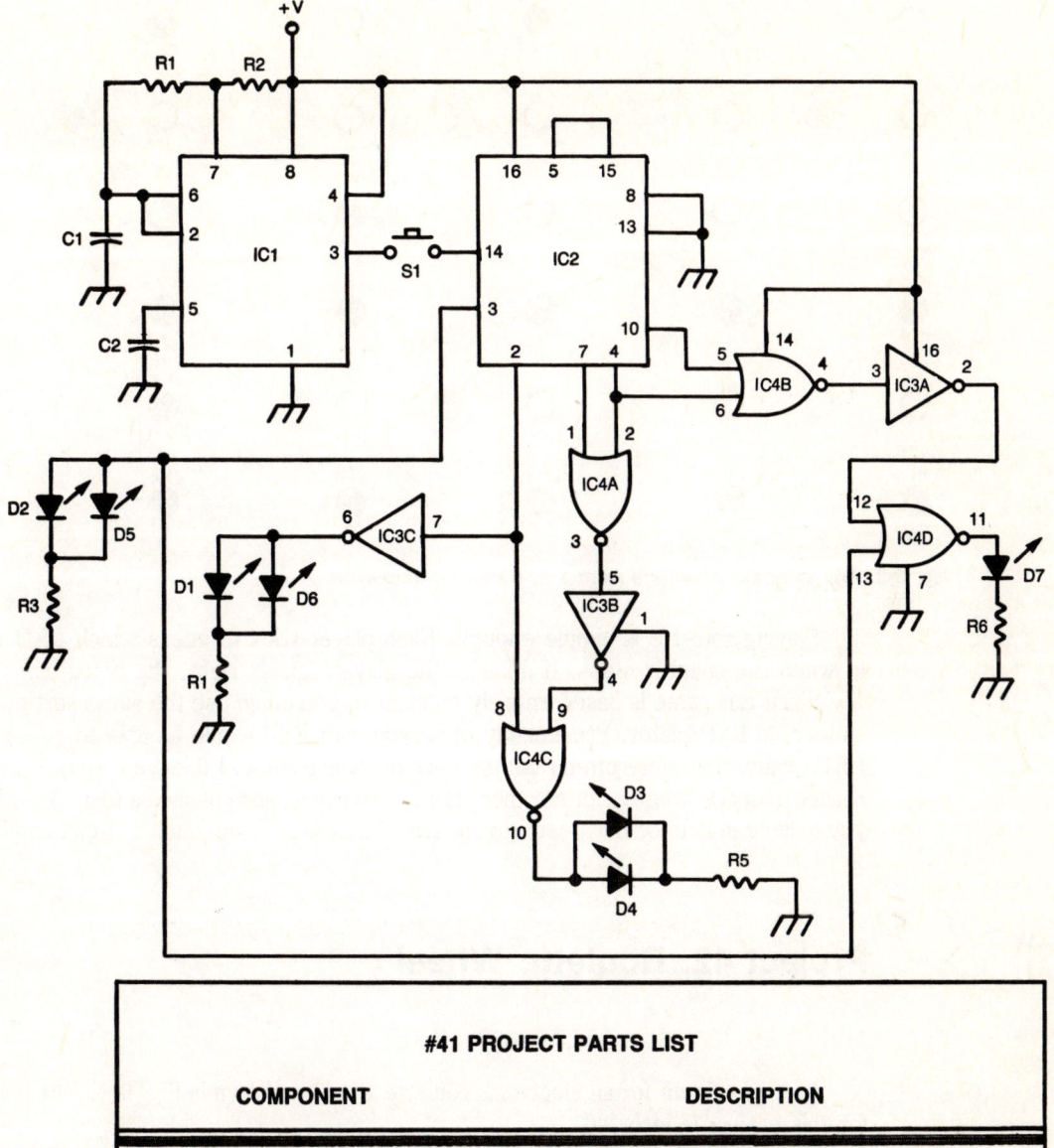

#41 PROJECT PARTS LIST

COMPONENT	DESCRIPTION
IC1	7555 (or 555) timer
IC2	CD4017 decade counter
IC3	CD4049 hex inverter
IC4	CD4001 quad NOR gate
D1 - D7	LED
C1, C2	0.01 μF capacitor
R1, R2	1kΩ resistor
R3 - R6	330Ω resistor
S1	SPST push switch (Normally Open)

161

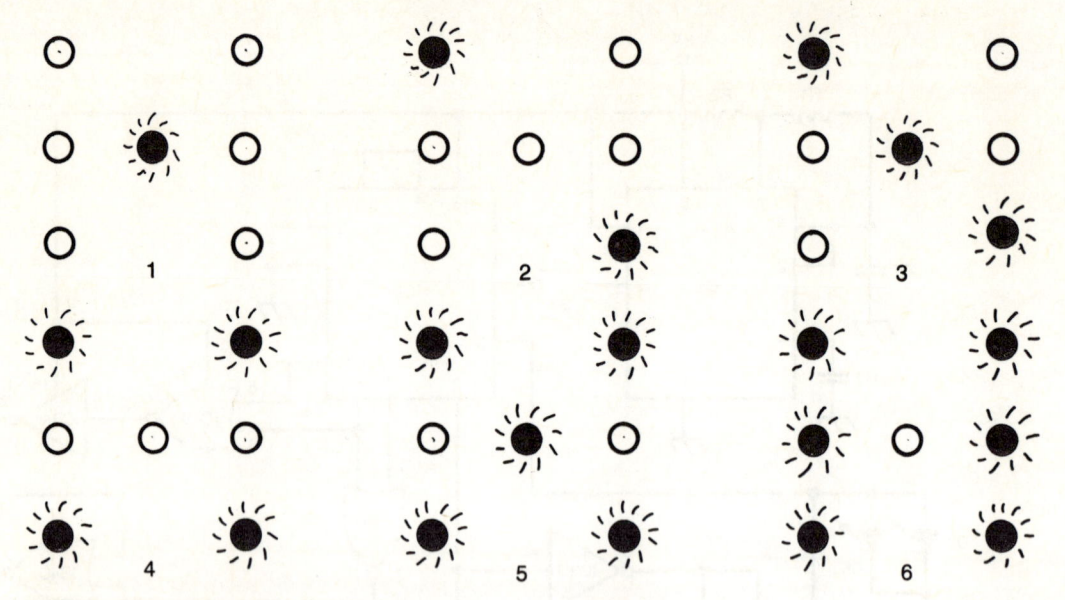

Fig. 9-3. Any standard dice face from 1 to 6 can be displayed.

Playing roulette is simple enough. Each player tries to guess which LED will be lit when the counter stops. If none of the players guess right, the house wins.

Since this game is based entirely on chance, you could use the same sort of circuit as a crude ESP tester. Presumably, a person with ESP would be able to guess the lit LED in advance, more often than the laws of chance allow. I'll leave it to the individual reader to decide whether or not there is any real significance of such a test. But whether you believe in ESP or not, testing your ESP "potential" with such a circuit can be fun.

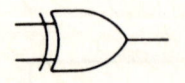

Project 42: Roulette Wheel

A simple circuit for an electronic roulette wheel is shown in Fig. 9-4. The parts list for this project is included.

This circuit is not very difficult to understand. It is made up of a simple binary random number generator that drives a 74C154 four-to-sixteen line decoder. For any given count only one of the sixteen outputs will be active, so only one of the sixteen LEDs will be lit.

While switch S1 is closed, the counter will cycle through its count sequence very rapidly, so all sixteen LEDs might appear to be continuously lit.

This simple circuit is functional, but it is not a very realistic simulation of a real roulette wheel. This seems to reduce a lot of the excitement from the game.

An improved electronic roulette wheel circuit appears in Fig. 9-5 with its parts list.

Fig. 9-4. A simple roulette wheel simulator.

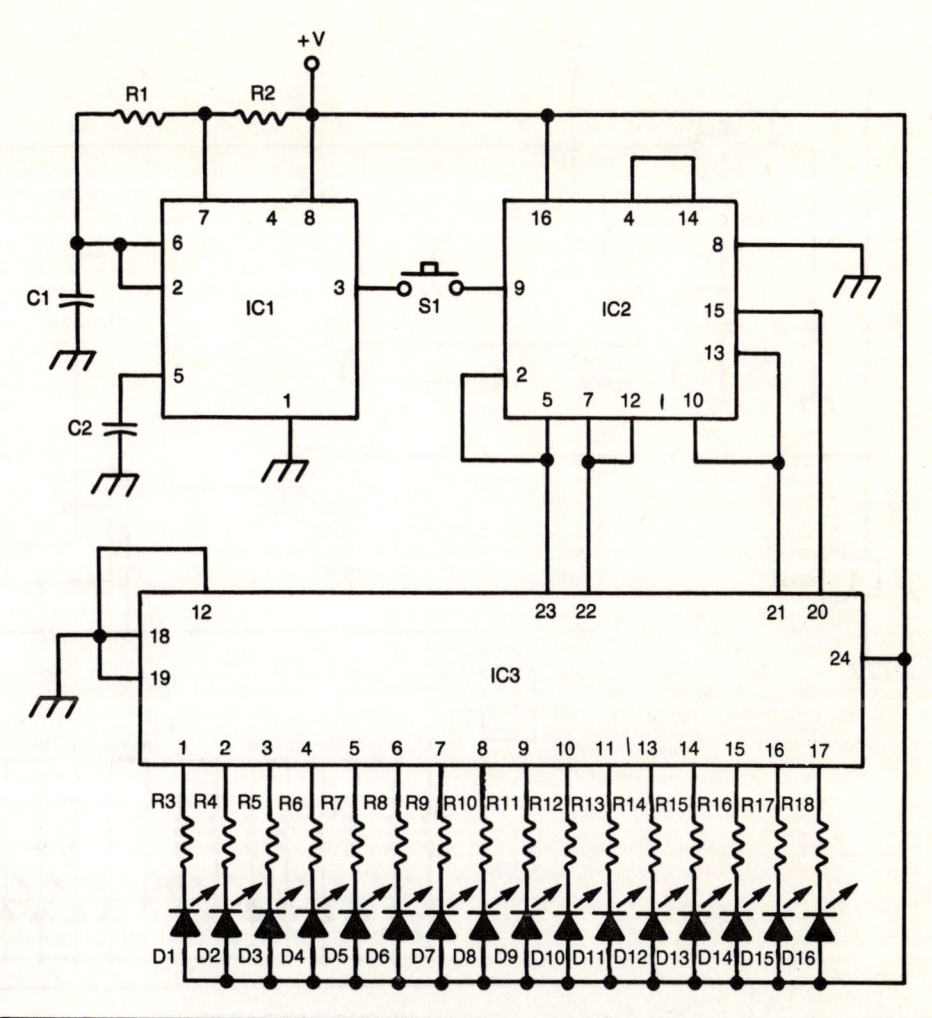

#42 PROJECT PARTS LIST

COMPONENT	DESCRIPTION
IC1	7555 timer
IC2	74C175 quad D flip-flop
IC3	74C154 4-to-16 line decoder
D1 – D16	LED
C1	0.1 µF disc capacitor
C2	0.01 µF disc capacitor
R1, R2	100kΩ resistor
R3 – R18	330Ω resistor
S1	SPST push-button switch (Normally Open)

#42 PROJECT PARTS LIST

COMPONENT	DESCRIPTION
IC1	CD4009 hex inverter
IC2	74C175 quad D flip-flop
IC3	74C154 4-to-16 line decoder
Q1	2N2222, or almost any low power NPN transistor
D1	1N4148 diode (or similar)
D2 - D17	LED
C1, C3	2.2 µF 25 volt electrolytic capcitor
C2	330 µF 25 volt electrolytic capacitor
R1, R2	120kΩ resistor
R3, R4	10kΩ resistor
R5	470kΩ resistor
R6 - R21	330Ω resistor
S1	SPST push-button switch (Normally Open)
SPKR	small speaker

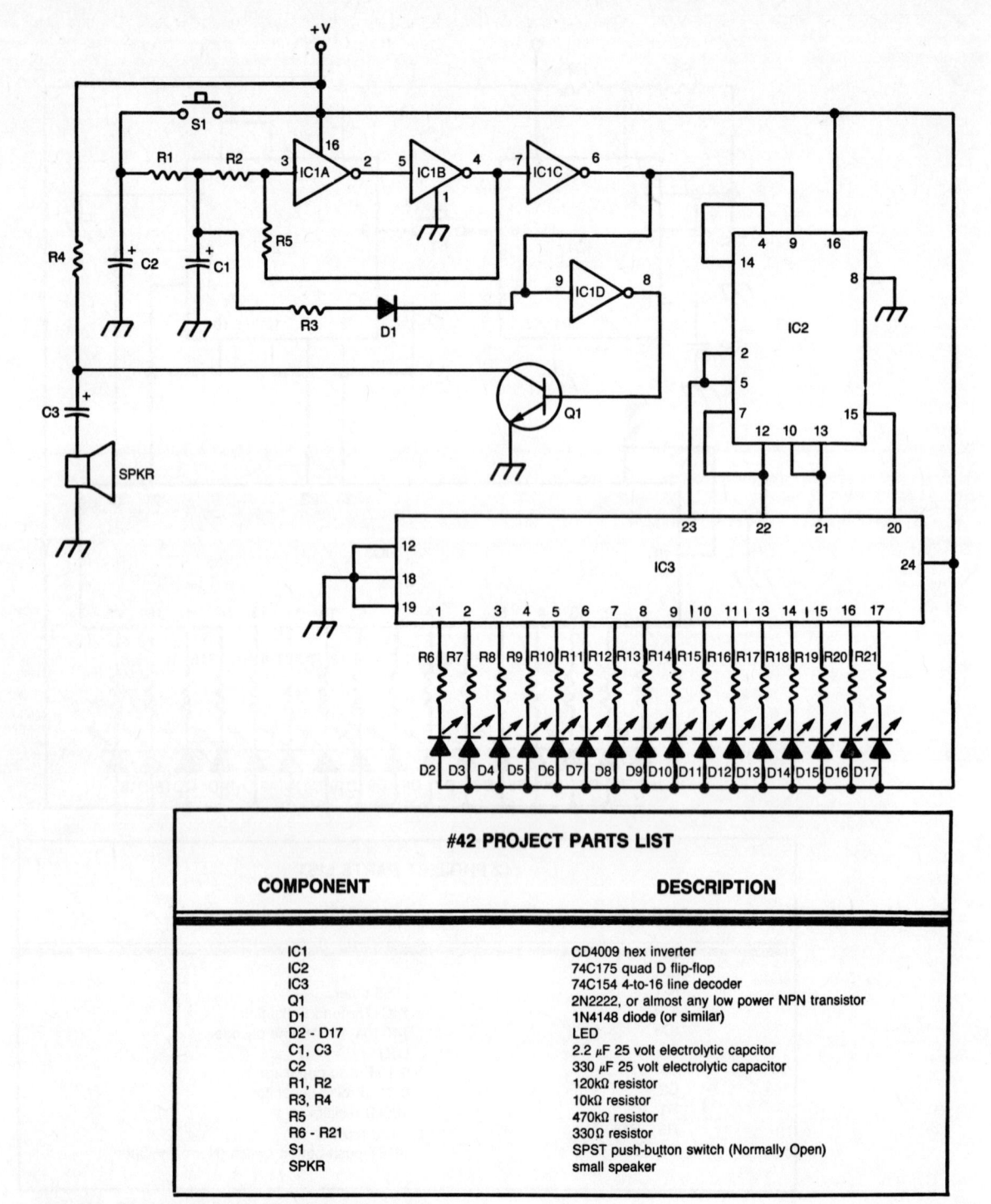

Fig. 9-5. An improved roulette wheel simulator.

In this version the simple 7555 timer-based clock is replaced with a clock formed from three inverters. When push-button switch S1 is momentarily closed, the circuit will start oscillating. The frequency will start out fairly high, so all the LEDs will appear to be lit. (Incidentally, the frequency can be altered by changing the value of capacitor C1). When the switch is released, the frequency will gradually slow down. At some point you will be able to see the individual LEDs lighting up in sequence, giving a definite sense of movement around a wheel. The movement will eventually come to a complete stop and only one LED will remain lit. The slow down rate is controlled by the value of capacitor C2. Increasing the capacitance of C2 will make the wheel "spin" longer. A smaller capacitor will result in a shorter sequence.

For even greater realism, a clicking sound is produced each time the count is incremented. If you don't want the sound effect, simply eliminate the following components from the circuit:

IC1D; Q1; R4; C3; SPKR.

If you want the sound effect sometimes, but don't want it every time you play, just put a SPST switch in one of the speaker lines.

Transistor Q1 is not at all critical. Almost any "garden variety" NPN transistor will work in this application.

For an effective roulette wheel, you will probably want to use a lower clock frequency than in a basic random number generator, or the electronic dice project. The roulette game is enhanced if we can see the wheel go round and round. If all the LEDs seem to be lit, the display won't look much like a spinning wheel.

For the first version with the fixed clock rate, the components listed in the parts list will give a frequency that is a little under 50 Hz. The frequency determining components in Fig. 9-4 are R1, R2, and C1.

The circuit of Fig. 9-5 is designed with a variable clock rate. It will start out with a very *fast* spin, and gradually slow down to a stop, just like a real (mechanical) roulette wheel.

10

MISCELLANEOUS CIRCUITS

In writing a book of electronic projects there is always several interesting circuits that don't fit neatly into any chapter heading. This final chapter is a potpourri of more or less unrelated projects using CMOS devices.

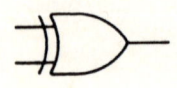

Project 43: Switch Debouncer

The common switch is probably the simplest component in most electronic circuits. It only has to make or break a connection, there obviously isn't much to it.

No mechanical switch is perfect, rather than neatly making firm contact when it is closed, a switch will tend to bounce open and shut several times very rapidly before settling into position. For the vast majority of analog applications this doesn't make the slightest bit of difference. The bounces are too short to be noticeable in the operation of the circuit.

On the other hand, digital circuits are designed to recognize very brief pulses, so bouncing switch contacts could be a real problem in many digital circuits. For instance, let's say we have a digital counter set up to keep track of the number of times a push button is pressed. Each bounce of the contacts would look like a separate switch closure to the counter. Instead of counting one push of the button as 1 (which is what we want), the circuit will count a random number of bouncing pulses. We expect to get:

0 - 1 - 2 - 3 - 4 - 5 - 6 - 7 - 8 - 9 - 0 -
1 - 2 - 3 - 4 - 5 - 6 - 7 - 8 - 9 - 0 - . . .

But what we end up getting is something like this:

0 - 6 - 1 - 4 - 0 - 3 - 8 - 2 - 7 - 1 -
6 - 2 - 5 - 9 - 3 - 7 - 0 - 4 . . .

Obviously, this kind of problem could well render a digital circuit completely useless.

What we need to prevent such problems is some way to get the digital circuitry to ignore the unwanted and meaningless bouncing contact pulses. A good way to accomplish this is to have the first closure of the contacts trigger a monostable multivibrator with an output pulse width that is longer than all the bouncing (settle-in time) of the switches contacts. When used this way the monostable multivibrator is known as a switch debouncer.

A suitable switch debouncer circuit is shown in Fig. 10-1. A parts list for this project is included with the figure.

As shown in the diagram, when the switch contacts first make contact, however briefly, the one-shot (made up of a pair of NOR gates) is triggered. Further switch openings and closings will simply be ignored until the multivibrator completes its timing period. This period does not have to be long (by human standards). A fraction of a second will be sufficient. In fact, if the delay time is made too long, some valid input pulses (switch closures) might be missed.

You should be aware that bounce-free switches will not be necessary for every manually operated switch in every digital circuit. But when the need does arise, as it inevitably will sooner or later, this simple circuit can make the difference between a functional piece of equipment, and a useless piece of high-tech junk.

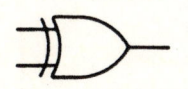

Project 44: Selectable Delay

In some applications, we will need something to happen after a specific delay period. A monostable multivibrator can be used for this purpose, but sometimes it has to be inverted. Sometimes extra circuitry is required in addition to the inversion.

To understand the problem, take a look at Fig. 10-2. Part A shows the normal output of a monostable multivibrator. The output is LOW until the circuit is triggered. The output then goes HIGH for the duration of the multivibrator's time period (set by component values within the circuit). After the circuit time period runs out, the output goes LOW again.

Fig. 10-1. A switch debouncer circuit.

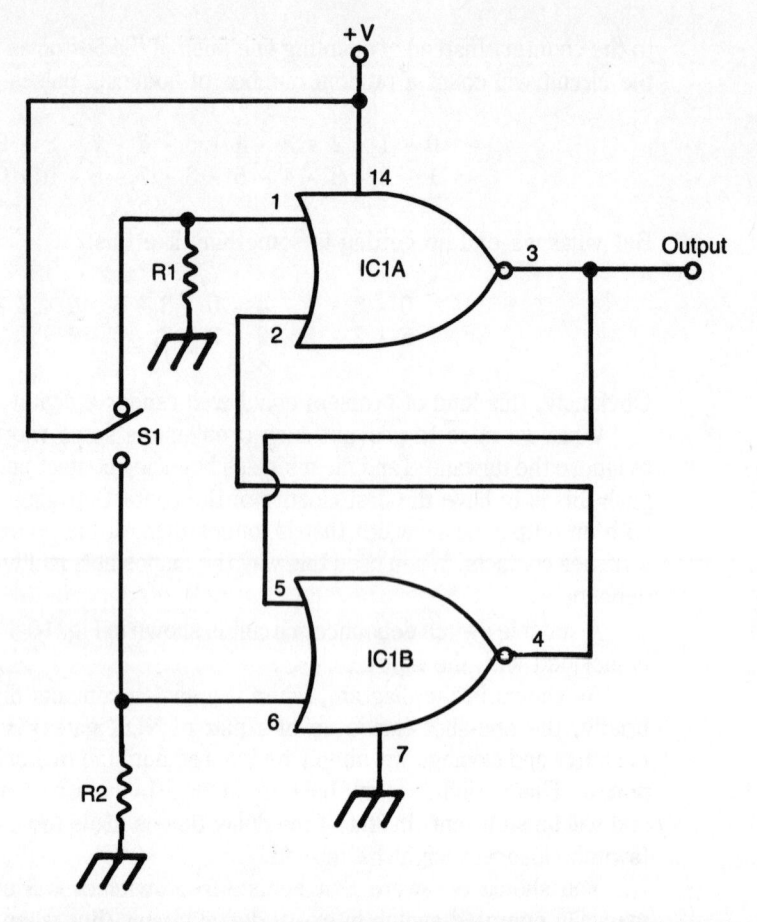

| #44 PROJECT PARTS LIST ||
COMPONENT	DESCRIPTION
IC1	CD4001 quad NOR gate
R1, R2	100kΩ resistor
S1	SPDT pushbutton

But what if we need the timer's output to go HIGH after a specific delay period? One simple solution is to invert the output from a standard monostable multivibrator. The inverted output is shown at B of Fig. 10-2. Now everything is simply reversed, the output will be normally HIGH, and will go LOW when the circuit is triggered. After the monostable's time period has run out the output will go HIGH again. This will work in some applications, but sometimes it will be problematic.

The problem often is the normally HIGH state. In some applications, the signal must be LOW, except for a pulse after the delay period. It should be LOW before the delay

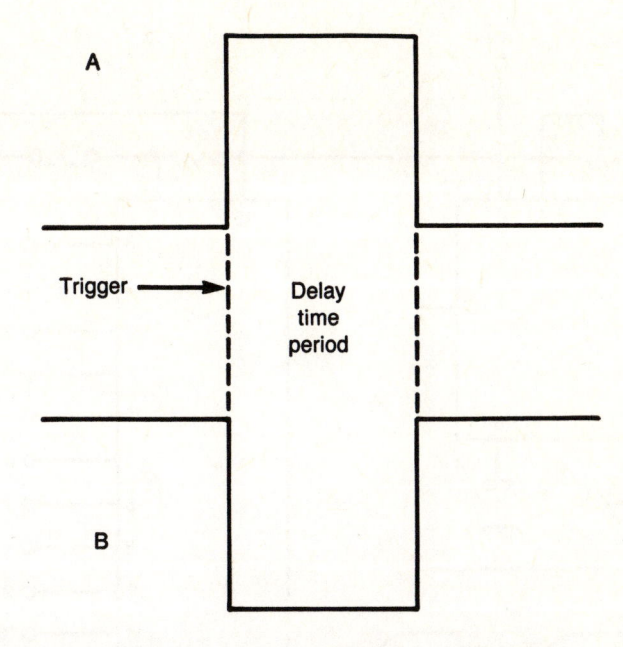

Fig. 10-2. Just inverting a delay pulse isn't sufficient for all applications.

period, and during the delay period. The desired signal is illustrated in Fig. 10-3. This could be achieved with a fairly complex gating network, but such a solution would be awkward, and inelegant at best.

I wouldn't start discussing such a problem unless I had a solution handy, the solution is to use a decimal counter, fed by a slow clock. The counter should be set up in the count-up-and stop mode. When the desired count is reached, the appropriate output goes HIGH. This output was LOW all the preceding time.

A fringe benefit of this technique, is that the time period is user selectable. Any multiple of the clock's basic time period (from 1 to 10) can be selected. For maximum versatility, use a 10PST rotary switch.

The complete schematic for this project is shown in Fig. 10-4 with the parts list.

This project is not terribly difficult to understand, if you break the schematic down into sections. Like many sophisticated projects this one is really just a combination of basic circuits.

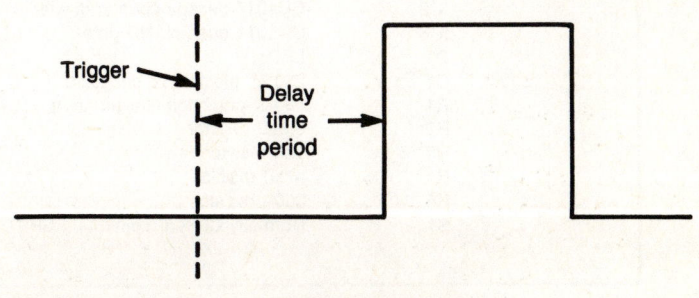

Fig. 10-3. A typical output from the selectable delay project.

Fig. 10-4. The selectable delay circuit.

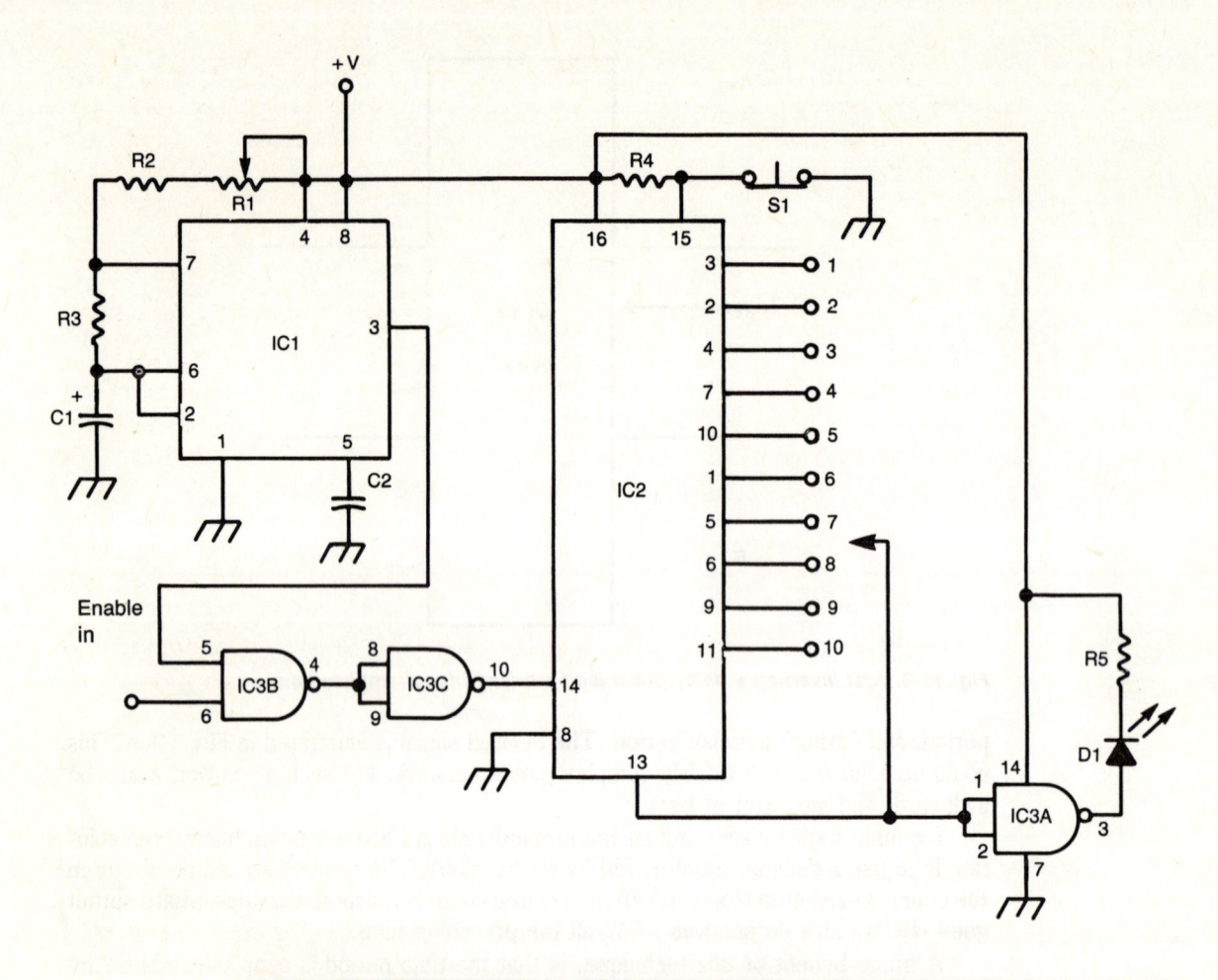

#44 PROJECT PARTS LIST	
COMPONENT	DESCRIPTION
IC1	7555 CMOS timer
IC2	CD4017 decade counter/divider
IC3	CD4011 quad NAND gate
D1	LED
C1	250 µF electrolytic capacitor
R1	500kΩ potentiometer (trimpot)
R2	6.8kΩ resistor
R3	1kΩ resistor
R4	47kΩ resistor
R5	330Ω resistor
S1	Normally Closed momentary SPST push switch

IC1 and its associated components serve as the low-frequency clock. IC1 is a 7555 CMOS timer in the astable mode. A standard 555 could be used, if you prefer.

Potentiometer R1 should be a trimpot type. This control allows you to adjust the basic time period within a fairly wide range. If you want to set a time outside R1's adjustment range, substitute a different capacitor for C1. The larger this capacitance is, the longer the basic time period will be. Using the component values shown in the parts list, the time period will be roughly in the one minute range.

The output from the clock is fed to an AND gate made up of two sections of IC3. The other input to the AND gate is an external ENABLE signal. If the ENABLE signal is LOW the clock's pulses will not reach the timer (IC2), and nothing will happen. When the gate is ENABLEd, the clock pulses can get through and advance the counter. In some applications, an external ENABLE signal might not be used. In this case eliminate IC3B and IC3C from the circuit and use a direct connection from IC1 pin #3 to IC2 pin #14.

IC2 is a CD4017 decoder counter in the count-and-halt mode. When the selected count is reached, the counter will stop advancing until it is reset. S1 is a manual reset switch. This should be a momentary action type switch (a push-button would be best). It must be a normally closed (NC) switch. Opening this switch briefly will reset the counter back to 0.

To select the desired multiple of the basic time period, just use the appropriate count pin as the output. Pin #13 must be connected to this pin too, to stop the counter. This output connection is indicated by the arrow in the schematic.

The output from the counter is fed to IC3A, a buffer/inverter. The LED will be lit when the output is HIGH. To use this circuit to drive some other device, just delete D1 and R5.

A simple variation on this project would be a continuously cycling version. After a delay equal to the basic time period multiplied by the selected clock value, the output will go HIGH for one time period, then it will automatically go LOW for another delay period. This will be repeated as long as the circuit is enabled. Changing the circuit for this mode of operation couldn't possibly be simpler. All you have to do is reverse the connections to pin #13, and pin #15 of IC2.

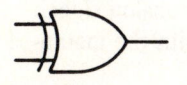

Project 45: Pulse Width Modulator

To transmit or record information, some form of modulation is often used. The information is called the *program*. The program is overlaid on a second signal called a *carrier*. Some aspect of the carrier signal is controlled or modulated by the program. The program forces the carrier to deviate from its nominal qualities.

In analog systems several forms of modulation are used. The most common are AM (amplitude modulation) and FM (frequency modulation). These forms of modulation are not difficult to figure out from their names. In AM, the amplitude, or instantaneous level of the carrier varies with the program. In FM, the carrier's frequency varies in step with the program. Other types of modulation also exist, such as phase modulation.

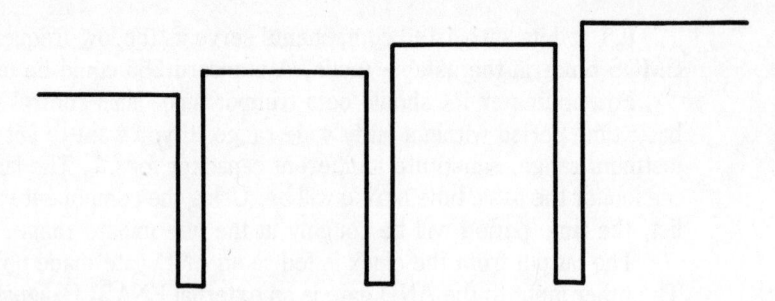

Fig. 10-5. Very narrow rectangle wave pulses.

In all types of modulation, the carrier usually has a much higher frequency than the highest frequency contained in the program.

One form of modulation blurs the distinction between analog and digital. This is PWM, or *pulse width modulation*. A rectangle wave can vary from a narrow pulse, as shown in Fig. 10-5, to a wide pulse illustrated in Fig. 10-6. In PWM, the actual width of the pulse is controlled by the program. This is illustrated in Fig. 10-7.

In a sense, PWM is a varied form of FM, since the pulse frequency is affected by changes in the pulse width.

Many different circuits can be used for pulse width modulation. One of the simplest is shown in Fig. 10-8. This circuit is built around a 7555 CMOS timer. Notice that there are two inputs, so that both the carrier and program signals can be fed in.

This circuit, unlike most using the 7555, does not include a stability capacitor. This is because pin #5 (voltage control input) is being actively used here.

Basically this circuit is a voltage-controlled monostable multivibrator circuit. The carrier signal is used to repeatedly trigger the multivibrator. If the carrier frequency is too high, the circuit will not work. If you run into problems here, decrease the values of R1 and C1. These components control the time constant of the circuit. A shorter time constant will permit higher frequencies to be used.

The time constant is not controlled solely by R1 and C1, however. These components just set the nominal value of the time constant, which is also affected by the voltage being fed into pin #5. This voltage is the program, which can be any analog signal.

The output will be a string of pulses whose output varies in step with the instantaneous amplitude of the program signal.

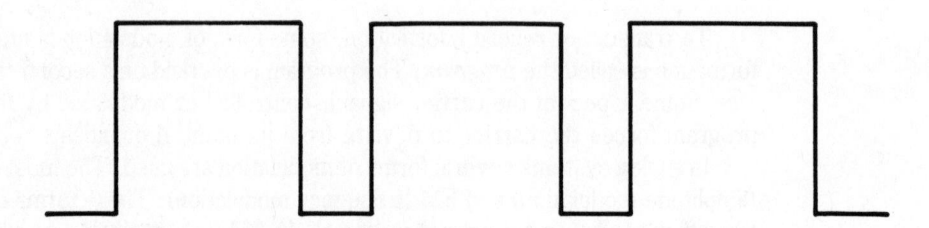

Fig. 10-6. Very wide rectangle wave pulses.

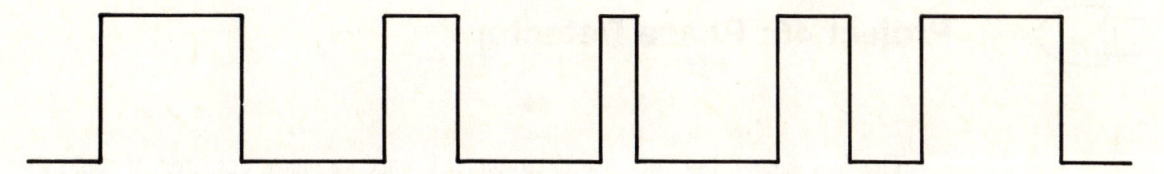

Fig. 10-7. A PWM signal features varying pulse widths.

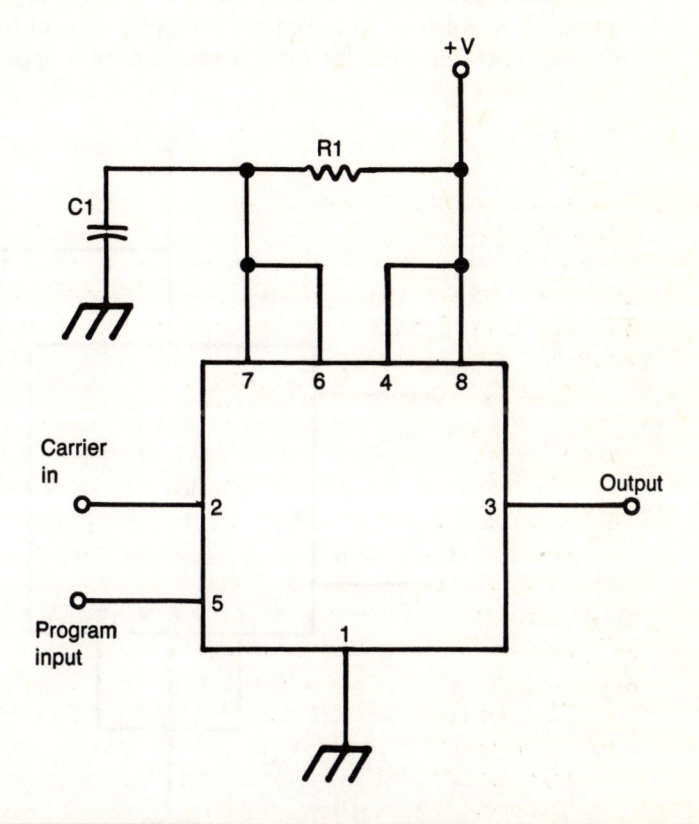

| #45 PROJECT PARTS LIST ||
COMPONENT	DESCRIPTION
IC1	7555 timer
C1	0.01 uF capacitor (*)
R1	10K resistor (*)

(*) experiment with other component values.

Fig. 10-8. A simple PWM circuit.

173

Project 46: Phase Detector

This is a simple, but very useful circuit, it is called a phase detector. It could also be called a "frequency comparator."

The basic phase detector circuit is illustrated in Fig. 10-9. As you can see, there really isn't much to it. It's really just a D-type flip-flop, with an LED to indicate the output state. Resistor R1 is a current-dropping resistor to protect the LED from excessive current. The parts list for this simple project appears with the figure.

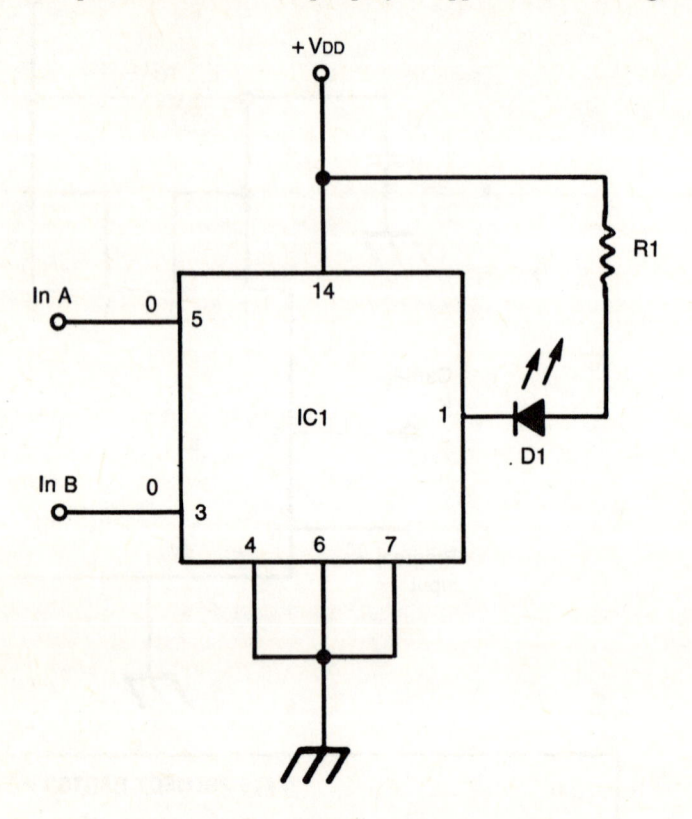

| #46 PROJECT PARTS LIST ||
COMPONENT	DESCRIPTION
IC1	CD4013 dual D-type flip-flop
D1	LED
R1	330Ω resistor

Fig. 10-9. This circuit indicates when two input signals are in phase.

Two input signals are fed into this circuit's inputs. If they are in phase with each other (cycles begin and end at the same times), the LED will light up. Otherwise, the LED will remain dark.

Many digital systems require synchronized signals. If clock signals are not reaching all relevant parts of the circuit simultaneously, incorrect operation can result. A complex digital system can be rendered absolutely useless by an out of sync signal. A phase detector gives a clear indication that two tested signals are in phase with each other.

Incidentally, the phase detector also tells you if the two signals are at the same frequency. This is why I said the circuit could be called a *frequency comparator*. If the two input signals are not operating at exactly the same frequency, they can briefly be in phase with each other, but they won't be able to stay in phase for long, since one signal by definition has shorter pulses than the other. So all of the pulses can't possibly begin and end in synchronization. The LED will light if the two signals are at the same frequency and in phase with each other.

This circuit can also respond to two signals that are harmonically related, if one signal is at a frequency that is a whole number multiple of the other signal's frequency. For example, if signal A is 500 Hz, and signal B is 1000 Hz, then B is at the second harmonic of A. (500 × 2 = 1000). For every complete pulse of signal B, signal A goes through two complete pulses. If both signals begin pulses at the same instant, the phase detector can indicate that they are in phase with each other, even though A puts out an extra pulse for each cycle of B. Whether or not the circuit presented here will operate reliably in the harmonic mode will depend in large part on the actual waveshape of the input signals. They must have very clean, sharp, rise and fall times. The synchronization of the two signals must be very tightly matched.

The LED will actually blink on and off when the inputs are harmonically in phase, but if the frequencies involved are high enough, the LED will appear to be continuously lit, although it will probably glow a little more dimly than usual.

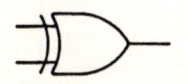

Project 47: Four Step Sequencer

In many applications we will need certain things to happen sequentially. A great many digital sequencer circuits have been developed for just this purpose. Most are based on some sort of counter being incremented by a clock.

A simple, but useful four step sequencer circuit is illustrated in Fig. 10-10 with its parts list.

IC1 is a 7555 (or 555) timer wired in the astable mode to serve as the system clock. The clock frequency can be adjusted via potentiometer R2, or by changing the value of capacitor C1. The larger this capacitor is, the lower the clock frequency will be. For this application we will generally want a fairly low clock frequency.

IC2 is a CD4013 dual flip-flop, this chip is wired as a two stage four step counter. The four NOR gates of IC3 create a one-of-four output pattern. Only one of the four outputs will be active (HIGH) at any time. The other three outputs will remain LOW.

For demonstration purposes, LEDs (D1 through D4) are used to indicate the output states. The LEDs will light up one by one in sequence. Only one LED should ever

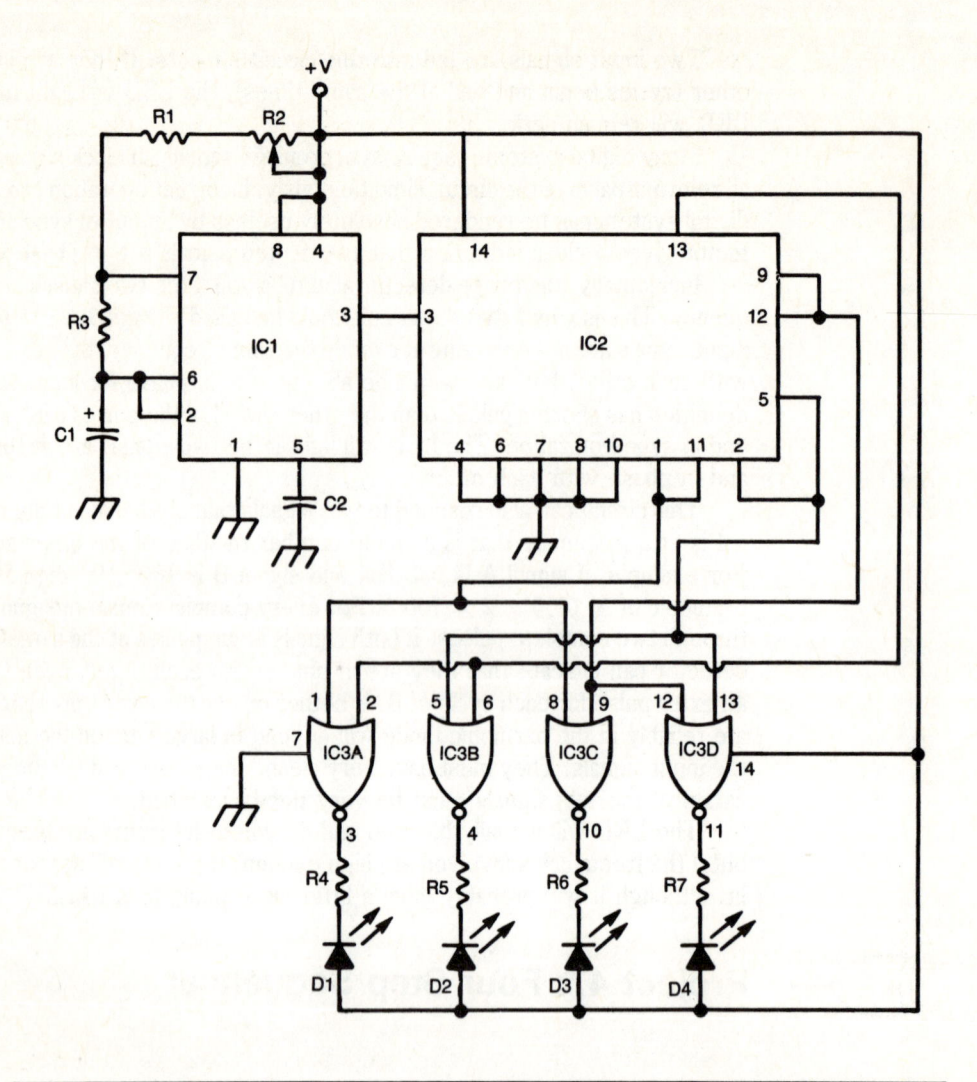

#47 PROJECT PARTS LIST	
COMPONENT	**DESCRIPTION**
IC1	7555 timer
IC2	CD4013 dual flip-flop
IC3	CD4001 quad NOR gate
D1 - D4	LED (*)
C1	25 µF electrolytic capacitor
C2	0.01 µF capacitor
R1, R3	1.2kΩ resistor
R2	500kΩ potentiometer
R4 - R7	330Ω resistor (*)

(*) = optional—see text

Fig. 10-10. A simple four step sequencer circuit.

be lit at any given instant. (If the clock frequency is too high, all four LEDs will appear to be simultaneously lit.)

If you are using this circuit to drive other circuitry, eliminate D1 through D4, and R4 through R7. Anything that can be controlled by a digital signal can be sequentially operated by this project.

Project 48: Frequency Multiplier

It isn't difficult to see how a signal's frequency can be divided by an integer value using flip-flops or counters. In some applications, we might need to multiply the signal frequency by an integer amount.

Digital frequency multiplication is a little trickier than digital frequency division, especially for odd integer values. It isn't too much of a problem to multiply the signal frequency by a power of two. This is done in the circuit of Fig. 10-11 along with its parts list.

As you can see, this is not a particularly complex circuit. It is made up of eight X-OR gates (two quad packages), and a pair of resistors. Actually, the resistor values aren't very critical. Use anything you happen to have handy, as long as it is reasonably close.

The circuit has two outputs, and a single input. At output A, the frequency will be equal to the input frequency multiplied by two. At output B, the input frequency is multiplied by four.

This circuit can easily be expanded with additional stages to multiply the input signal by any power of two. The third stage would multiply the frequency by eight, the fourth by sixteen, and so forth.

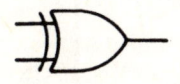

Project 49: Frequency Divider

It is often necessary (or at least desirable) to use signals with different frequencies in different parts of a system.

You could use multiple signal sources (clocks), but it might be unnecessarily expensive, more importantly, it might not even work in many cases. The two signal sources will almost certainly be out of phase with each other, making system-wide synchronization impossible. There are methods for locking the phase of one signal source to another, but such techniques tend to be rather complicated.

Obviously, I wouldn't have brought the problem up if there wasn't a handy solution available. The answer is to use a frequency divider. The original signal source is divided by a fixed value, resulting in a second (lower) frequency, that is in step with the original signal.

The frequency division techniques described here are for division by integer values only, that is, the divisor must be a whole number, such as 2 or 5. Fractional values, such as 1.5, cannot be used, at least not directly. There are ways to "cheat." Start out with a higher frequency that can be divided by two different integers to result in

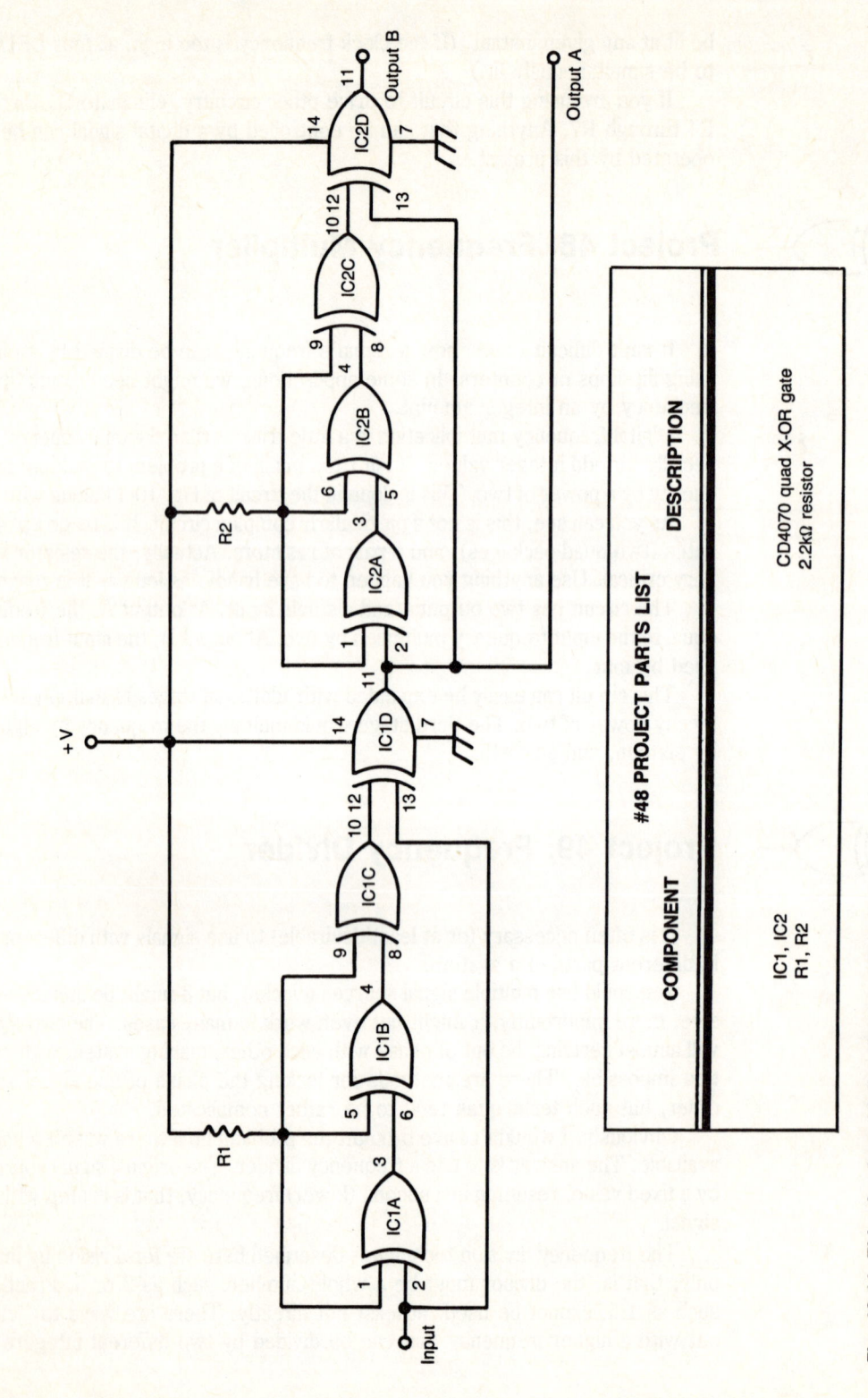

Fig. 10-11. This circuit multiplies the input frequency.

#48 PROJECT PARTS LIST

COMPONENT	DESCRIPTION
IC1, IC2	CD4070 quad X-OR gate
R1, R2	2.2kΩ resistor

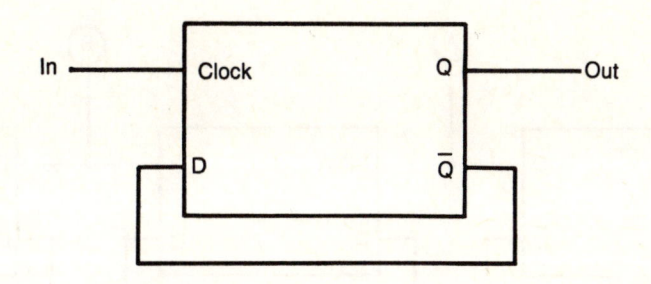

Fig. 10-12. This circuit divides the input frequency by two.

two signals with the desired fractional frequency relationship. For example, let's start out with a 1000 Hz signal, and divide it by 2 and 3. The first sub-signal (divide by 2) has a frequency of 500 Hz. The second sub-signal (divide by 3) has a frequency of 333.33 Hz. This gives us the same results we'd get if we started out with the 500 Hz signal and divided by 1.5 for the second frequency.

Digital frequency division is almost always accomplished through the use of flip-flops or counters (which are made up of cascaded flip-flops). A single flip-flop, as shown in Fig. 10-12, will act as a divide-by-two circuit. Remember that a flip-flop reverses its output state each time an input pulse is received. Let's assume that the flip-flop's output is initially LOW, as shown in Table 10-1:

INPUT PULSE #	OUTPUT STATE	OUTPUT PULSE #
0	LOW	0
1	HIGH	1
2	LOW	1
3	HIGH	2
4	LOW	2
5	HIGH	3
6	LOW	3
7	HIGH	4
8	LOW	4
9	HIGH	5

Table 10-1.

For every two input pulses, there is just one output pulse. The output frequency is one-half the input frequency.

By adding more flip-flops in sequence, we can divide by higher values. Each additional flip-flop raises the dividing value by a power of two. It is easy enough to see why. Consider the three-stage divider circuit of Fig. 10-13. Flip-flop A divides the original input signal by two. The output of this stage is F/2. Then flip-flop B divides the output of flip-flop A by 2. The frequency is halved again, so the output of the second stage has a frequency of F/4. This frequency is again divided by 2 by flip-flop C, resulting in an output of F/8. This can be continued indefinitely.

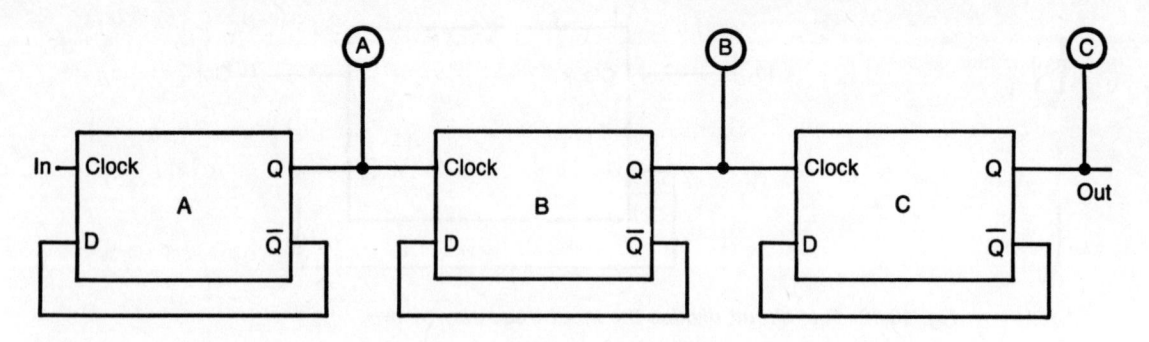

Fig. 10-13. Multiple stage frequency dividers divide input frequencies by higher values.

Cascading flip-flops in this manner creates a counter circuit. In a sense, the circuit counts the input pulses. When the count has reached eight, the circuit produces an output pulse, then starts over.

Suppose we need to divide the frequency by an integer that is not a power of two? The solution is simple enough. Just force the counter to reset itself prematurely when the desired value is reached. As examples, a divide-by-three counter is shown in Fig. 10-14. Figure 10-15 illustrates a divide-by-five counter circuit.

It is generally more convenient to use dedicated counter chips rather than separate flip-flops, but the principle is the same.

Figure 10-16 shows the schematic for a practical frequency divider circuit. The rotary switch (S1) selects the value the input frequency will be divided by. This circuit permits division by any integer from 1 to 10. (It may seem pointless to include divide-by-one, but it doesn't add to the circuit's complexity, and it makes things easy if you ever need the original undivided frequency.)

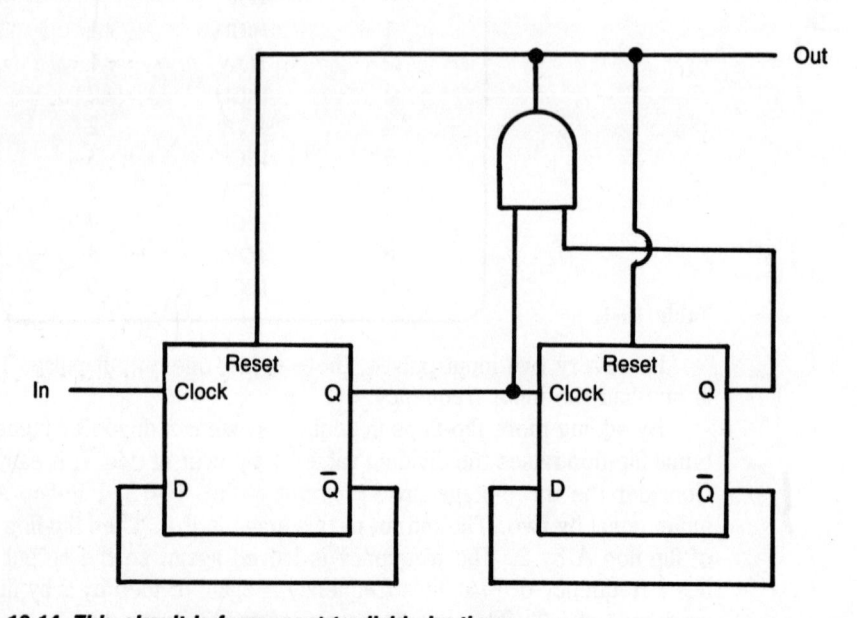

Fig. 10-14. This circuit is force-reset to divide by three.

180

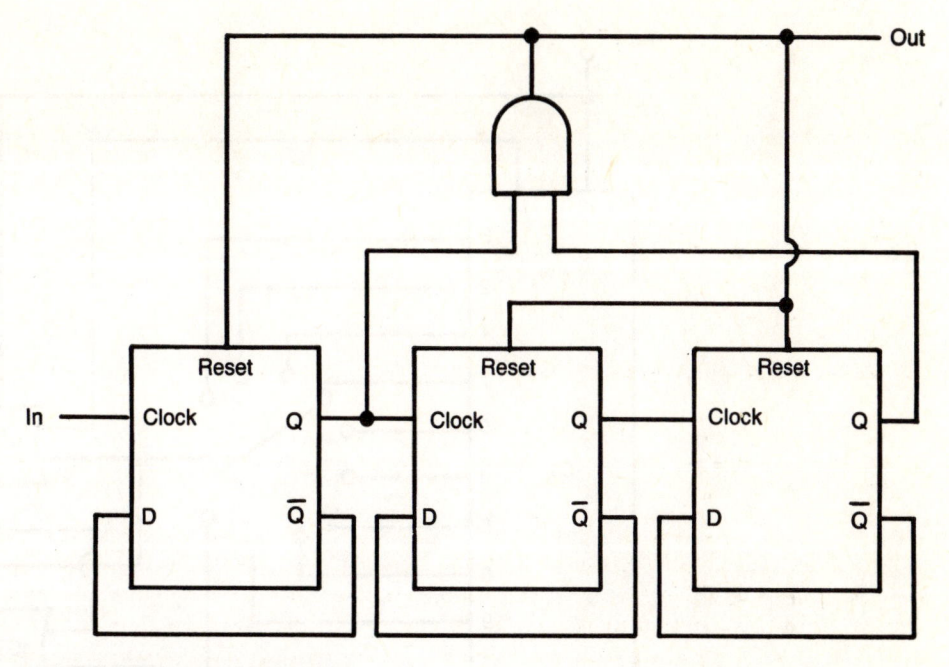

Fig. 10-15. A divide-by-five circuit.

It is easy to demonstrate the action of this circuit with a typical example. Let's assume that the input signal is a 35 kHz (35,000 Hz) square wave. The output for each position of switch S1 will be as shown in Table 10-2.

The major limitation of this type of circuit is that the duty cycle of the output signal will not be the same as that of the input signal in many cases. The output will always be a square wave with a duty cycle of 50 percent, regardless of the input signal's duty cycle. This won't matter at all in many circuits, but occasionally it will be a problem so you should be aware of it.

This circuit is very versatile, but surprisingly simple. As the parts list indicates, only three components are required.

FREQUENCY /DIVISOR	RESULTANT FREQUENCY	FREQUENCY /DIVISOR	RESULTANT FREQUENCY
F/1 =	35000 Hz	F/6 =	5833 Hz
F/2 =	17500 Hz	F/7 =	5000 Hz
F/3 =	11667 Hz	F/8 =	4375 Hz
F/4 =	8750 Hz	F/9 =	3889 Hz
F/5 =	7000 Hz	F/10 =	3500 Hz

Table 10-2.

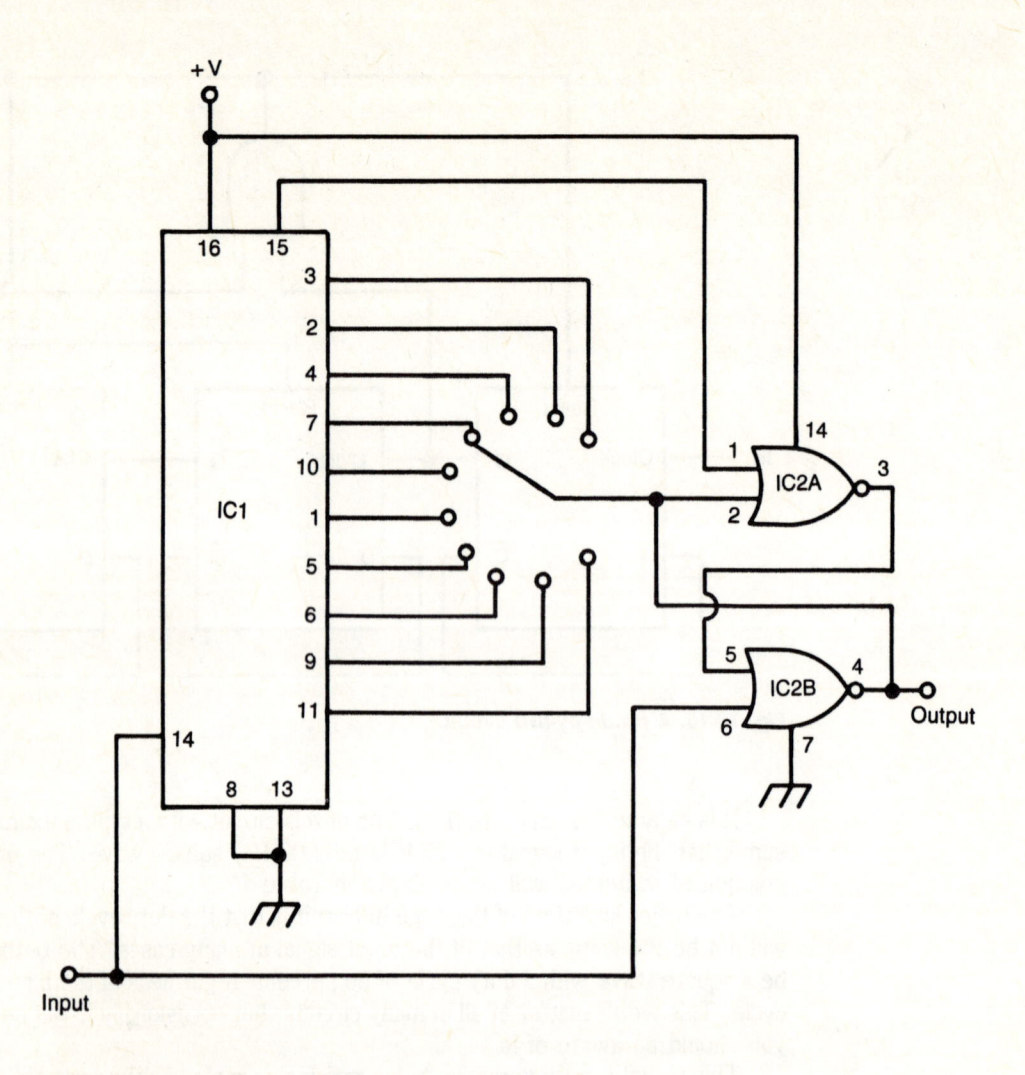

Fig. 10-16. A frequency divider with switch selectable division values.

#49 PROJECT PARTS LIST	
COMPONENT	DESCRIPTION
IC1	CD4017 counter
IC2	CD4001 quad NOR gate
S1	SP10T rotary switch

182

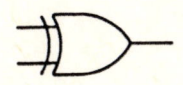

Project 50: Linear Amplifier

This project is certainly an odd one for a collection of digital circuits. It's a linear amplifier, made from a CMOS inverter. This is one of those circuits that you feel shouldn't have any right working, but somehow it does.

The basic circuit is shown in Fig. 10-17 with a typical parts list for this project. As you can see, there really isn't much to this circuit. Capacitor C1 blocks out any dc component in the input signal. The two resistors provide a feedback path, and determine the gain, according to this simple formula:

$$Av = R2/R1$$

For the component values given in the parts list, the gain works out to:

$$Av = 10,000,000/470,000 = 21.3$$

You might want to use a potentiometer and a fixed resistor in series for R2 to obtain manually variable gain.

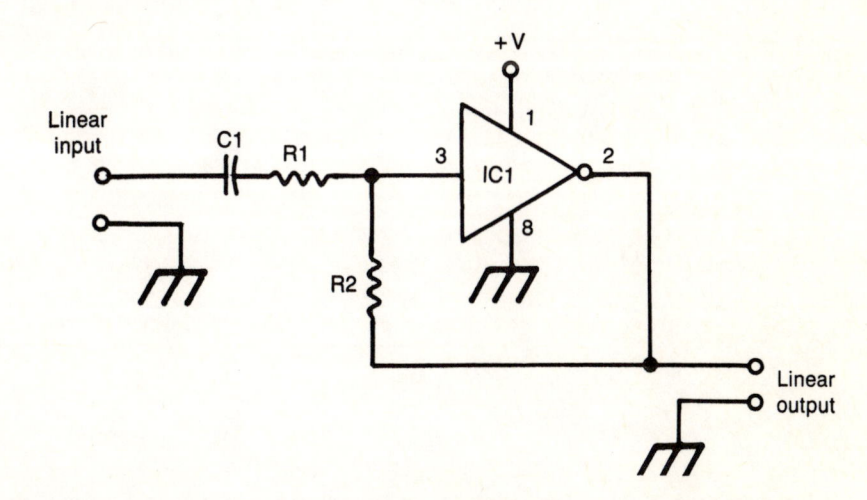

#50 PROJECT PARTS LIST

COMPONENT	DESCRIPTION
IC1	CD4049 hex inverter
C1	0.05 µF capacitor
R1	470kΩ resistor *
R2	10 MΩ resistor *

* (see text)

Fig. 10-17. A linear amplifier.

APPENDIX

APPENDIX

Appendix
Some Popular CMOS ICs

CD4001

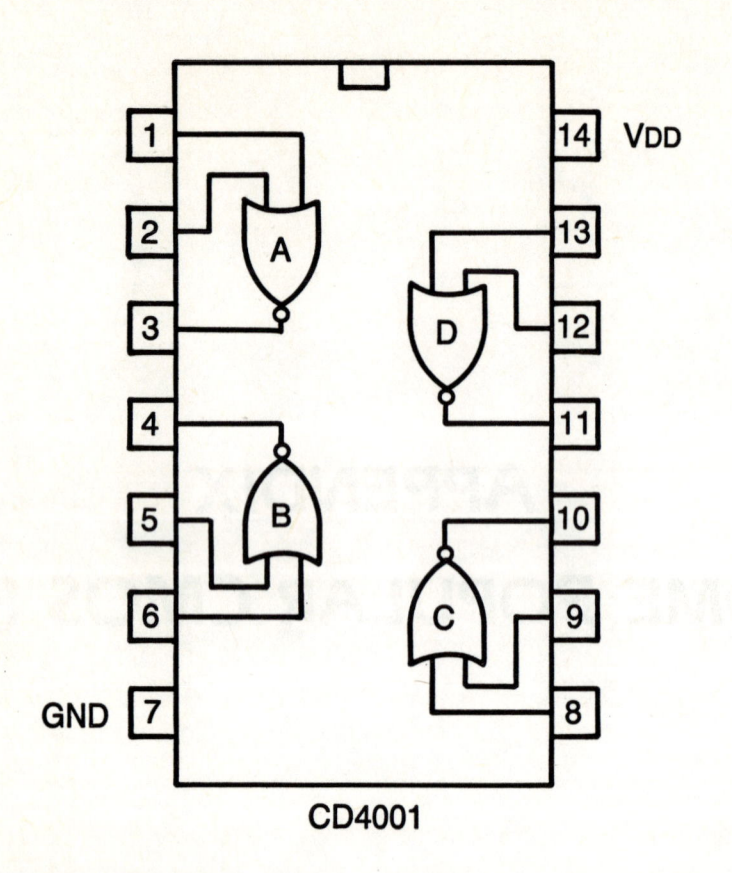

1			14	+VDD
2	A		13	
3			12	D
4			11	
5	B		10	
6			9	C
GND 7			8	

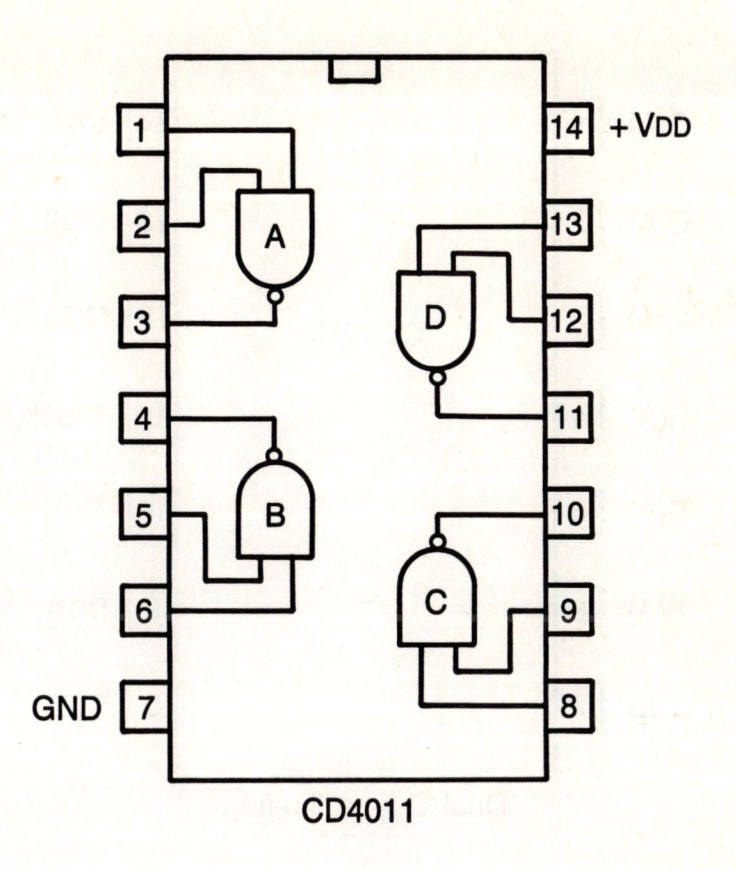

CD4011

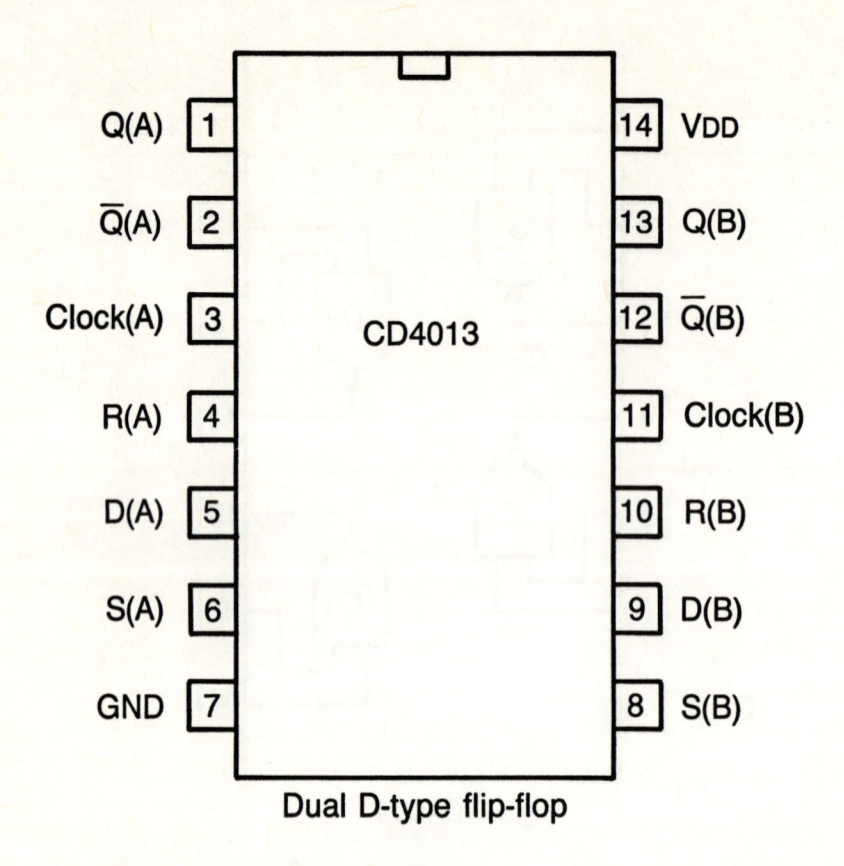

```
Q(A)      1              14   VDD
Q̄(A)      2              13   Q(B)
Clock(A)  3    CD4013    12   Q̄(B)
R(A)      4              11   Clock(B)
D(A)      5              10   R(B)
S(A)      6               9   D(B)
GND       7               8   S(B)
```

Dual D-type flip-flop

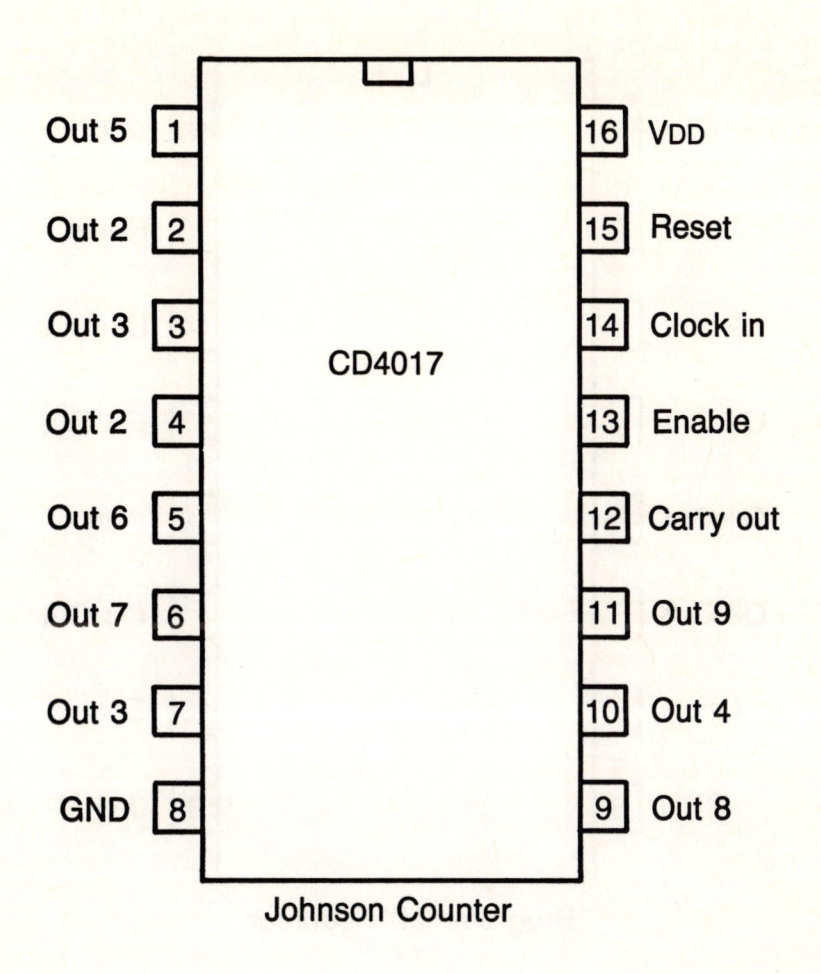

```
Out 5    1              16   VDD

Out 2    2              15   Reset

Out 3    3              14   Clock in
            CD4017
Out 2    4              13   Enable

Out 6    5              12   Carry out

Out 7    6              11   Out 9

Out 3    7              10   Out 4

GND      8               9   Out 8
```

Johnson Counter

191

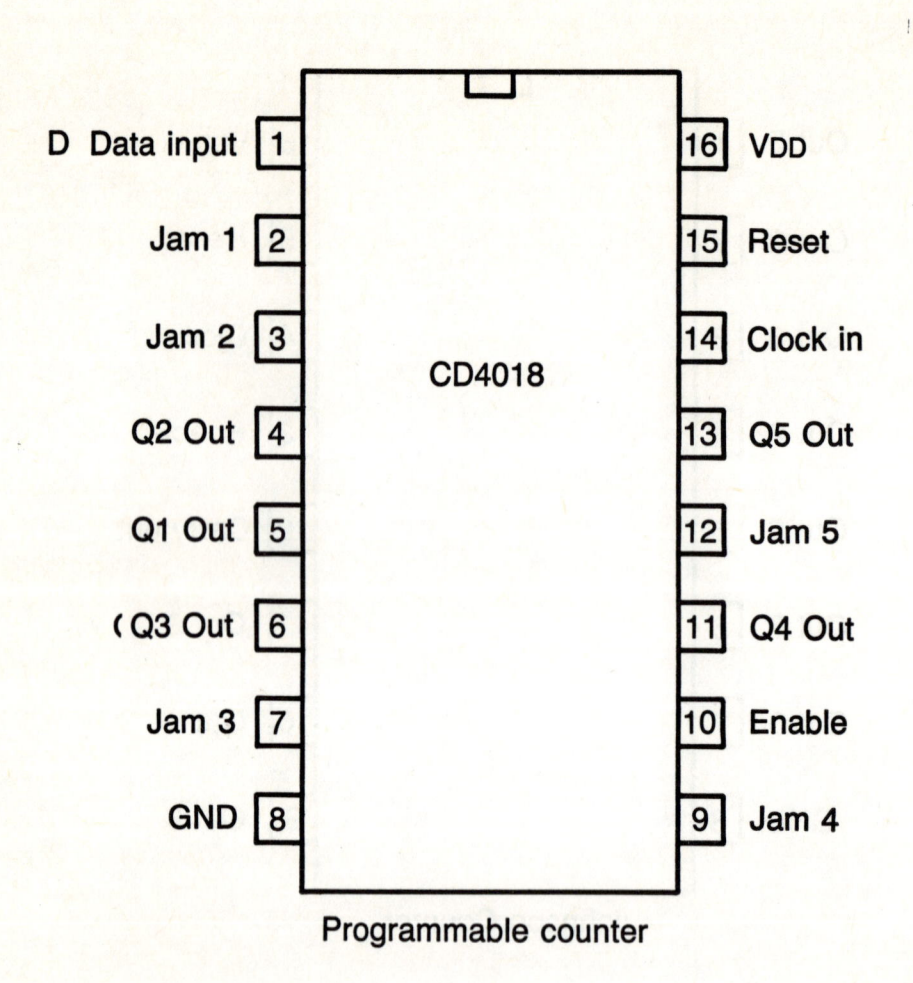

D Data input	1	16	VDD
Jam 1	2	15	Reset
Jam 2	3	14	Clock in
Q2 Out	4	13	Q5 Out
Q1 Out	5	12	Jam 5
Q3 Out	6	11	Q4 Out
Jam 3	7	10	Enable
GND	8	9	Jam 4

CD4018

Programmable counter

Triple three-input NAND gate

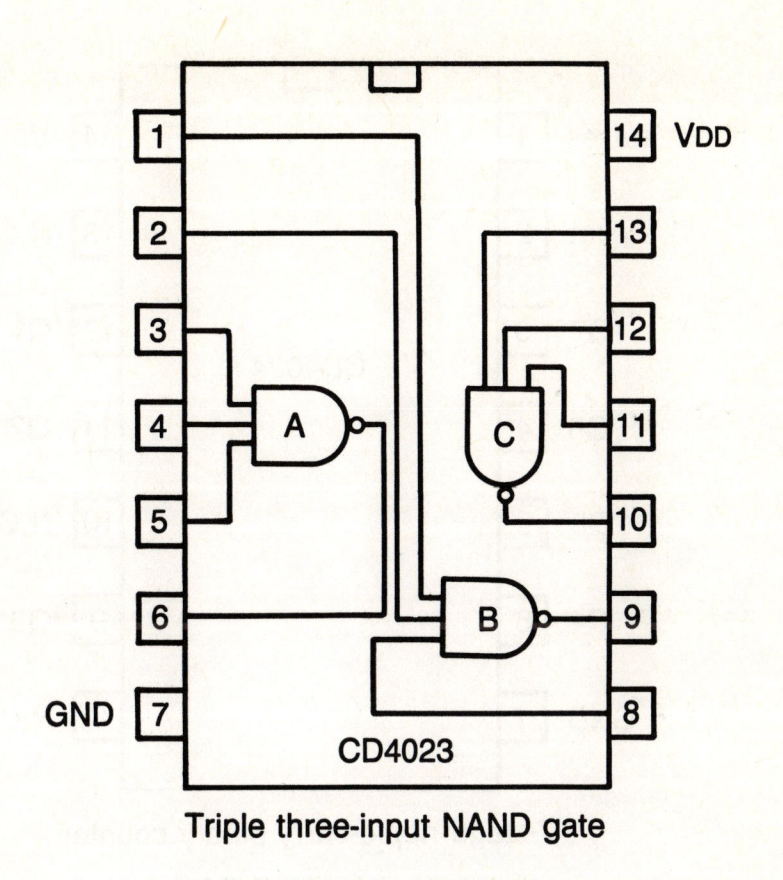

CD4023

Pin 7: GND
Pin 14: VDD

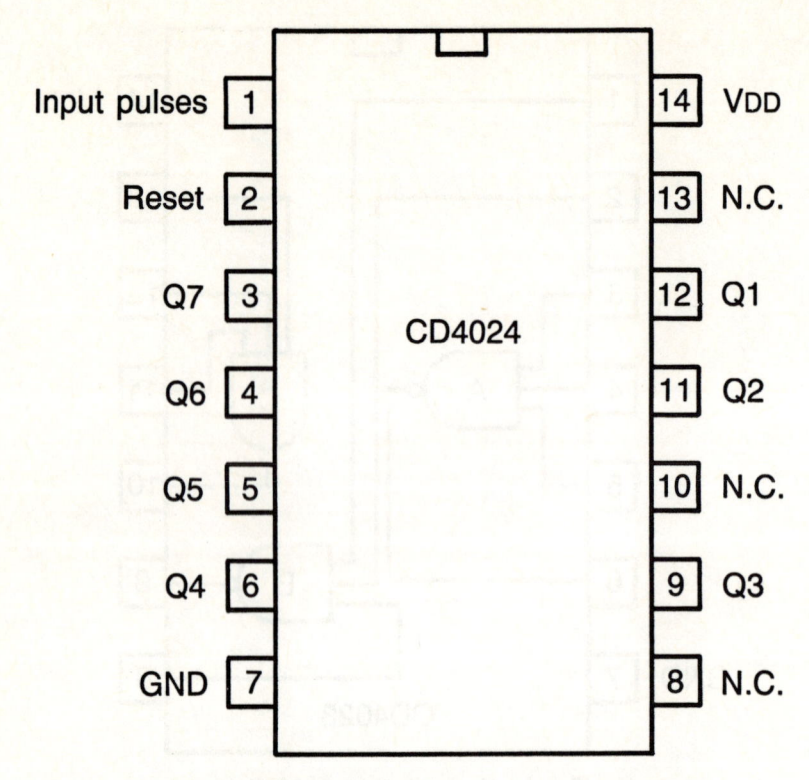

```
Input pulses  1           14  VDD
Reset         2           13  N.C.
Q7            3           12  Q1
                CD4024
Q6            4           11  Q2
Q5            5           10  N.C.
Q4            6            9  Q3
GND           7            8  N.C.
```

7-stage ripple-carry binary counter

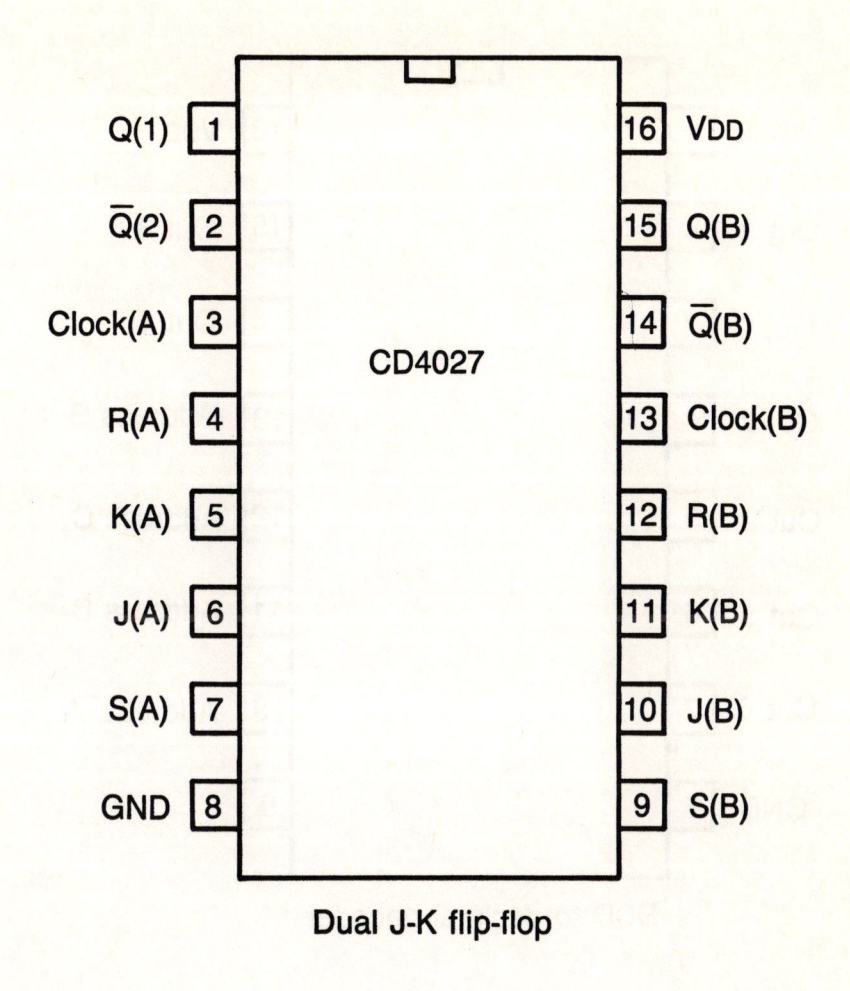

```
            ┌─────────┐
    Q(1)  1 │         │ 16  VDD
    Q̄(2)  2 │         │ 15  Q(B)
 Clock(A) 3 │ CD4027  │ 14  Q̄(B)
    R(A)  4 │         │ 13  Clock(B)
    K(A)  5 │         │ 12  R(B)
    J(A)  6 │         │ 11  K(B)
    S(A)  7 │         │ 10  J(B)
    GND   8 │         │  9  S(B)
            └─────────┘
```

Dual J-K flip-flop

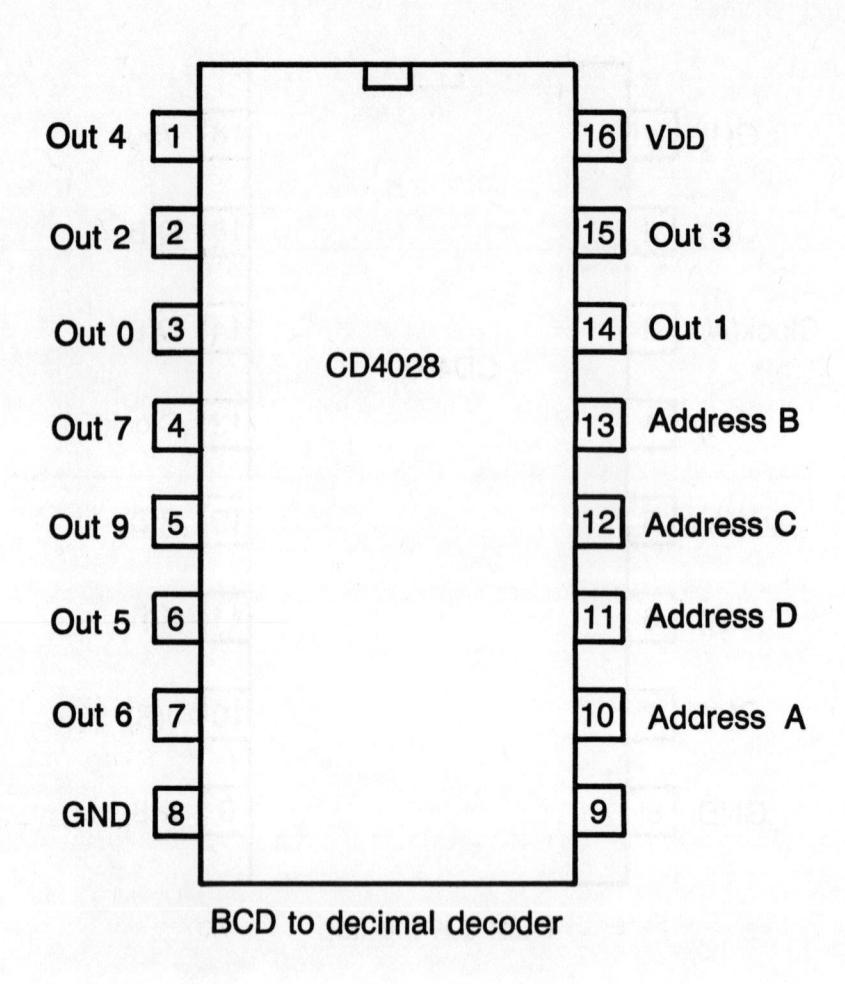

	CD4028	
Out 4 — 1		16 — VDD
Out 2 — 2		15 — Out 3
Out 0 — 3		14 — Out 1
Out 7 — 4		13 — Address B
Out 9 — 5		12 — Address C
Out 5 — 6		11 — Address D
Out 6 — 7		10 — Address A
GND — 8		9

BCD to decimal decoder

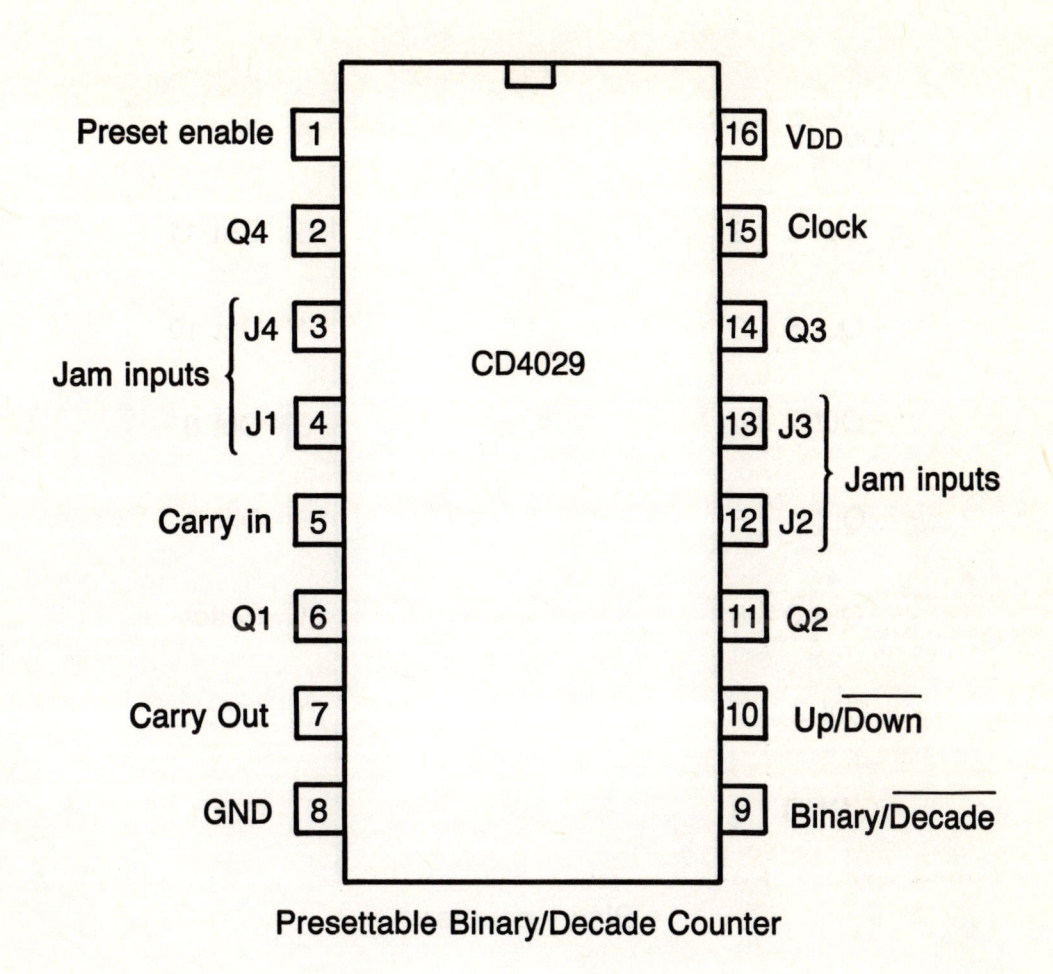

Preset enable	1	16	VDD
Q4	2	15	Clock
J4	3	14	Q3
J1	4	13	J3
Carry in	5	12	J2
Q1	6	11	Q2
Carry Out	7	10	$\overline{\text{Up/Down}}$
GND	8	9	$\overline{\text{Binary/Decade}}$

CD4029

Jam inputs: J4, J1 (pins 3, 4); J3, J2 (pins 13, 12)

Presettable Binary/Decade Counter

```
Out 12  [1]              [16] VDD
Out 6   [2]              [15] Out 11
Out 5   [3]   CD4040     [14] Out 10
Out 7   [4]              [13] Out 8
Out 4   [5]              [12] Out 9
Out 3   [6]              [11] Reset
Out 2   [7]              [10] Clock
GND     [8]              [9]  Out 1
```

Binary counter

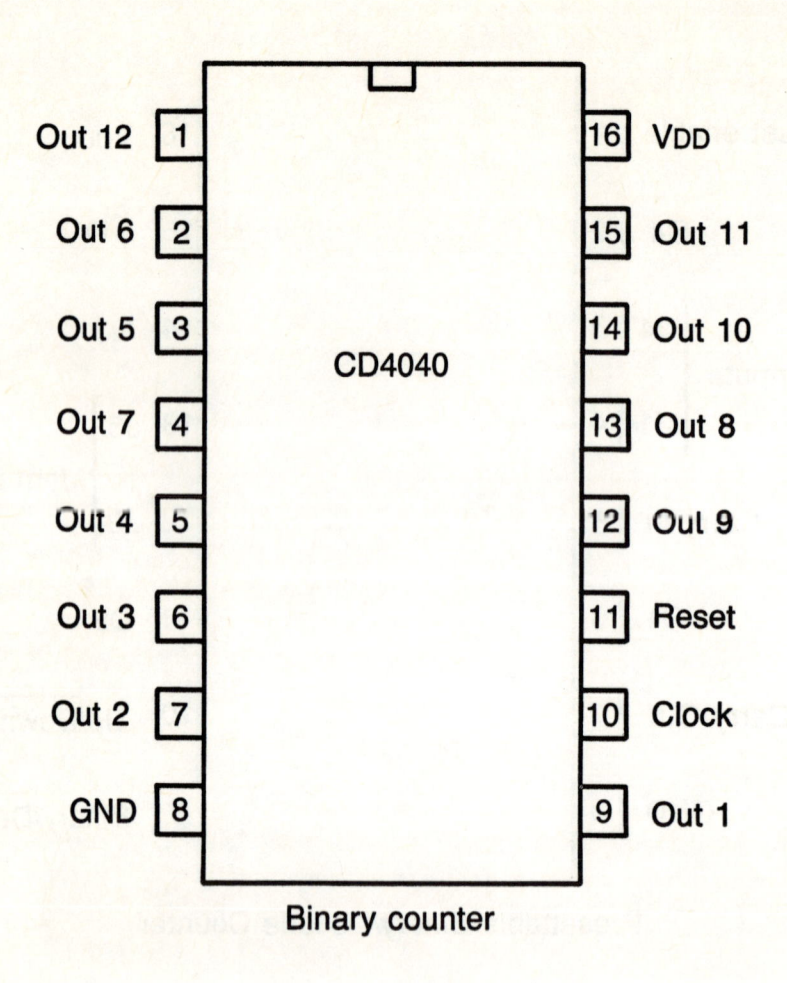

CD4049

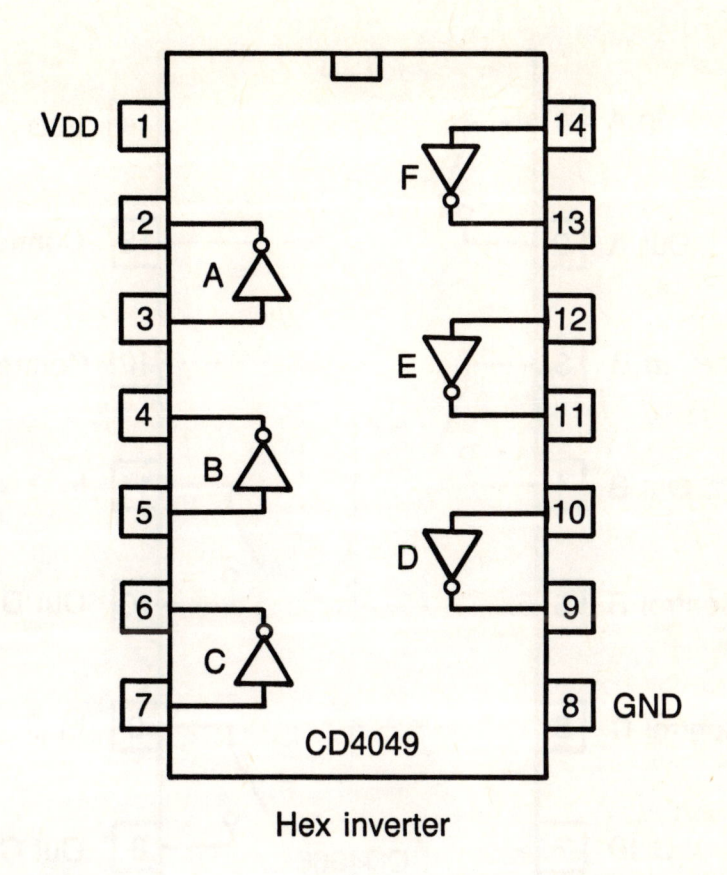

Hex inverter

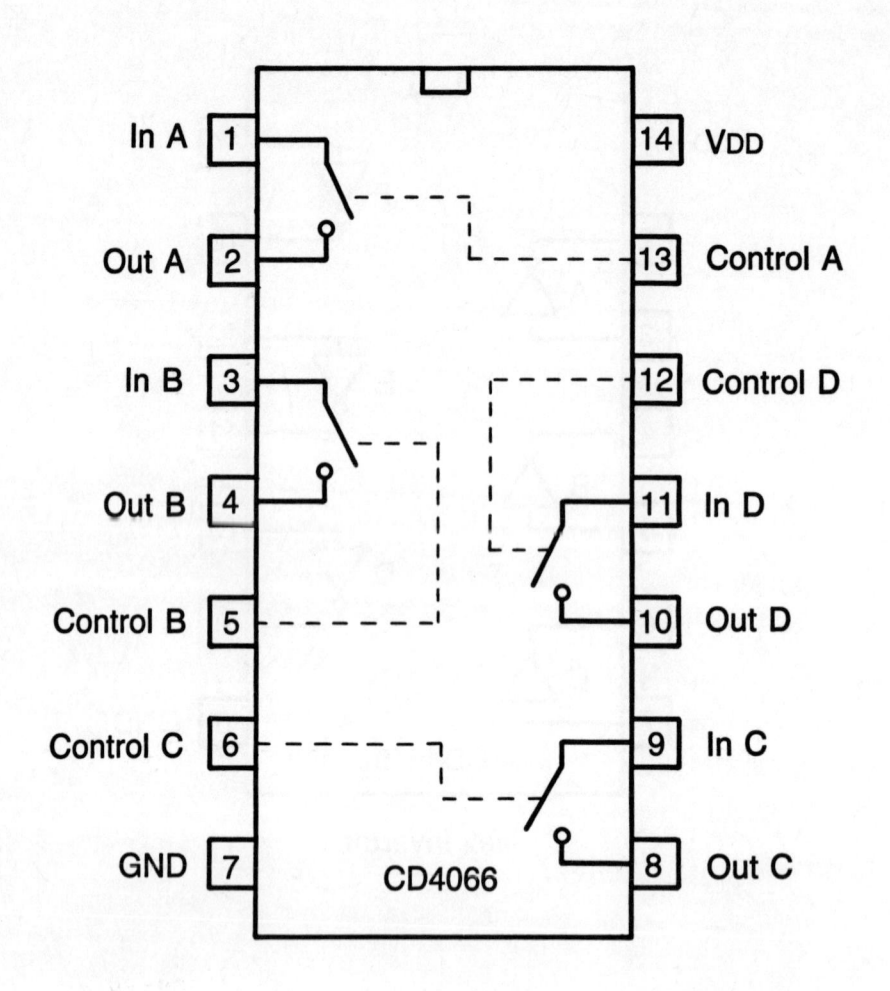

In A	1		14	VDD
Out A	2		13	Control A
In B	3		12	Control D
Out B	4		11	In D
Control B	5		10	Out D
Control C	6		9	In C
GND	7	CD4066	8	Out C

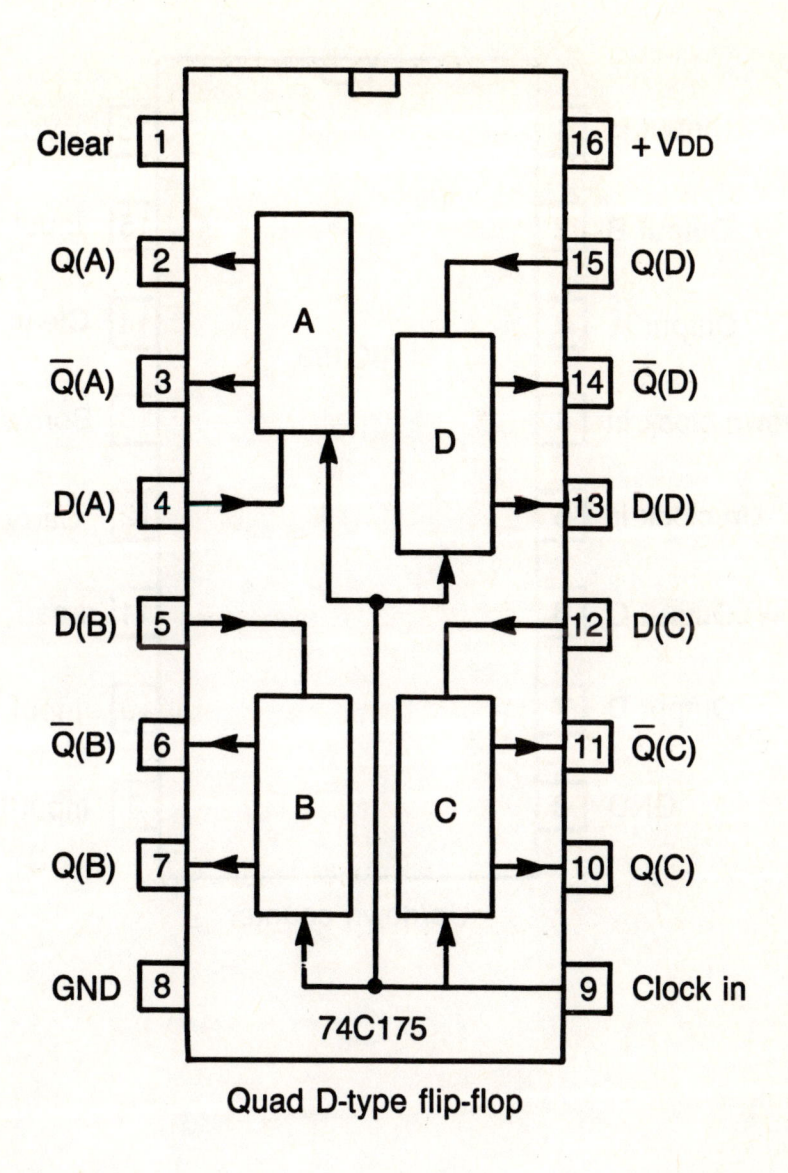

Pin	Label
1	Clear
2	Q(A)
3	$\overline{Q}$(A)
4	D(A)
5	D(B)
6	$\overline{Q}$(B)
7	Q(B)
8	GND
9	Clock in
10	Q(C)
11	$\overline{Q}$(C)
12	D(C)
13	D(D)
14	$\overline{Q}$(D)
15	Q(D)
16	+VDD

74C175

Quad D-type flip-flop

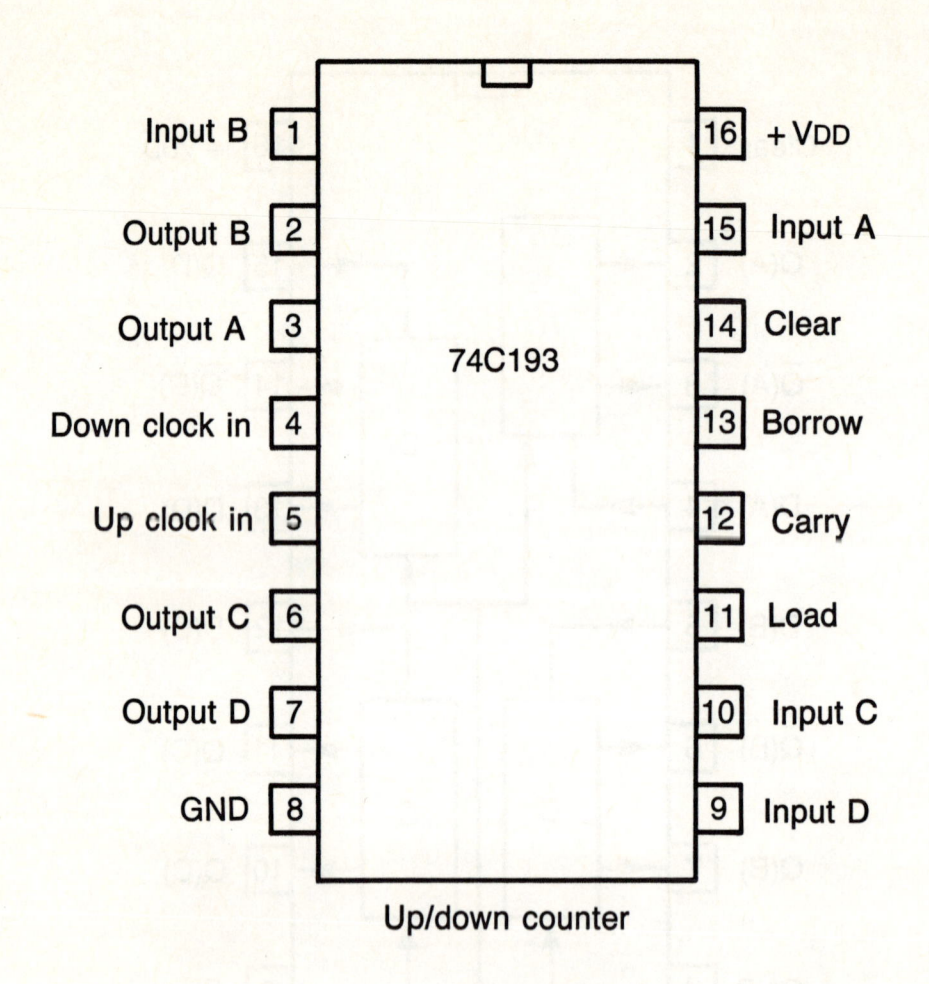

```
Input B        1 |          | 16   +VDD
Output B       2 |          | 15   Input A
Output A       3 |  74C193  | 14   Clear
Down clock in  4 |          | 13   Borrow
Up clock in    5 |          | 12   Carry
Output C       6 |          | 11   Load
Output D       7 |          | 10   Input C
GND            8 |          |  9   Input D
```

Up/down counter

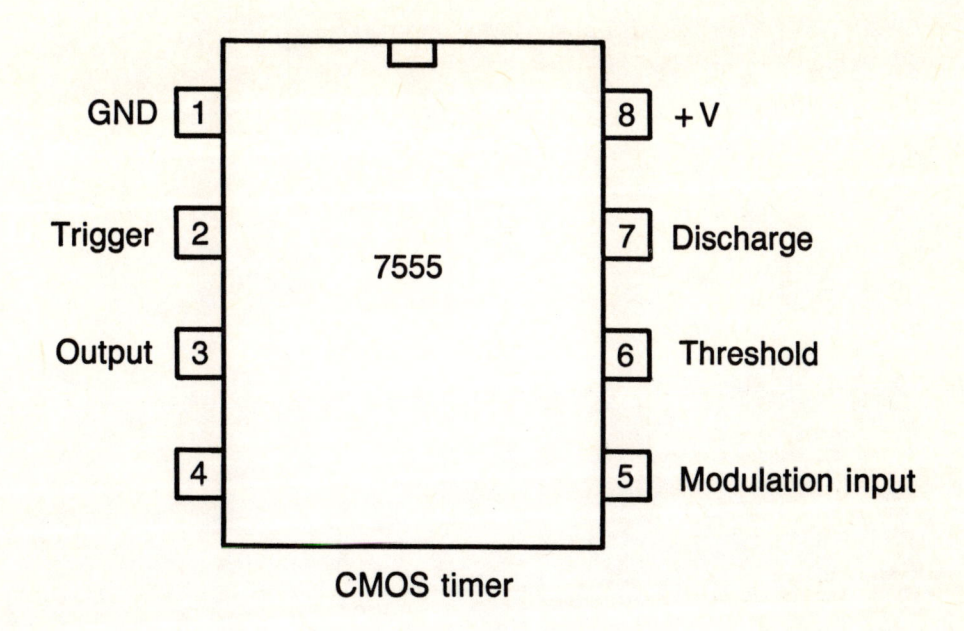

```
GND    [1]            [8]  + V

Trigger [2]           [7]  Discharge
              7555
Output [3]            [6]  Threshold

       [4]            [5]  Modulation input
```

CMOS timer

Index

Index

A
analog signals, 2
AND gate, 5, 6, 9, 10
appliance controller, 52
astable multivibrator, 12, 13
automatic night light, 50

B
band-pass filter, frequency response for, 112
band-reject filter, 112
BCD coding, 117
BCD-to-decimal decoder, 196
bidirectional switch, 105
bilateral switches, 53, 105
binary adder, 35-36
binary addition, 33-36
binary circuits, 23-41
   binary addition in, 33
   digital comparators as, 23
   majority logic circuits as, 37-41
   shift registers as, 28
binary counter, 198
binary numbering system, 34
binary-coded decimal (BCD) coding, 117
bistable multivibrator, 12, 14
bits, 2, 34
breadboarding, vii
buffer gate, 3
bytes, 2

C
capacitance
   programmable, 56
   variable, 53
capacitance meter circuits, 71-75
   schematic for, 74
carrier, 171
cascading flip flops, 180
clear input, 17
clock circuits, 149
clock input, 16, 135
   splitting, 83
CMOS
   definition of, 1
   input signal restrictions for, 21
   power supplies for, 21
   special precautions for, 20
   uses for, 20
CMOS timer, 203
comb filter, 113, 114
commuting filter, 113, 114
continuous mode, 126
control circuits, 42-58
   touch switches as, 42-53
   variable resistance/capacitance, 53
count-down timer, 129

counter circuits, 117-149
   flip flops used in, 18
counters, 17-20
cut-off frequency, 110, 111

D
D flip flop, 17
   dual, 190
   quad, 201
debouncer, 145
decimal count-down timer, 129
decimal output counter, 119
delay pulse, inverting, 169
difference detector, 8
digital clock, 155
digital comparator, 23-26
   multi-bit, 26
digital electronics, basics of, 1
digital filter, 110-116
   harmonics in, 113
   instructions for, 114-116
digital frequency meter
   block diagram for, 68
   fundamentals of, 67
   instructions for, 69-71
   signals within, 69
digital oscillator, 85
digital relay driver, 57-58
digital signals, 1

207

analog signals vs., 2
   transition time in, 2
digital sine wave generator, 92
diode-transistor logic (DTL), 20
disallowed state, 15
divide-by-five circuit, 181
dual D-type flip flop, 190
dual J-K flip flop, 195

E
electrolytic capacitor, 45
electronic dice, 159
enable input, 15

F
fan-out, 3
filters, digital, 110-116
flip flops, 12
   basic binary counter using, 123
   cascading, 180
   dual D-type, 190
   dual J-K, 195
   multi-digit counters using, 122
   quad D-type, 201
four-step sequencer, 174
four-tone sequencer, 100
frequency comparator, 174
frequency divider, 177
   multiple stage, 180
frequency meter, simple, 63
frequency multiplier, 177

G
game circuits, 159-165
gated oscillator, 85, 100, 101
gating networks, 11

H
harmonics, 113
hex inverter, 199
high-pass filter, 110, 111

I
input signal restrictions, 21
inverter gate, 4
inverters, 8
   triangle-wave generator of, 90

J
J-K flip flop, 17
   dual, 195
jam inputs, 92
Johnson counter, 191

L
LED display, 41
LED flashers, 76-83
   instructions for, 77-81
light activated gate, 46
linear amplifier, 183
logic families, 20

logic gates, 3
   majority logic circuits and, 37
   networks using, 11
   non-standard pattern combinations using, 9
   truth table for, 12
logic probe, 59
   deluxe, 64
   enable input for, 62
   pulse stretcher use of, 62
low-pass filter, 110

M
magnetic reed switch counter, 139
majority logic circuits, 37-41
majority logic demonstrator, 38-41
miscellaneous circuits, 166-183, 166
monostable multivibrator, 12, 13
   capacitance meter circuits using, 71
multi-bit digital comparator, 26
multi-digit counters, 122-148
multi-digit decimal counter, 126
multivibrators, 12
music-making projects, 84-116

N
NAND gates, 5, 6, 8, 10, 11, 193
night light, automatic, 50
nonequality gate, 8
NOR gates, 5, 8, 10
   R-S flip flop from, 15
notch filter, 112

O
one-hertz timebase, 151
one-minute timer, 152
   audible alarm for, 154
one-shot, 71
OR gates, 5, 7, 11
oscillators
   digital, 85
   gated, 85
   tunable, 86
oscilloscopes, 59

P
phase detector, 174
phase locked loops (PLLs), 86
phase relationships, 68
photo-theremin, 98
photoelectric counter, 133
   circuit for, 138
photoelectric object counter, 136
photoelectric rotation counter, 137
photosensitive sensor, 50
PISO shift registers, 28
power supplies, 21
preset input, 17
presettable binary/decade counter, 197
program, 171
programmable counter, 192

programmable resistance/capacitance, 53
pseudo-random flasher, 80, 81
pulse delayer, 47
pulse stretcher, 62
pulse width modulation, 172
   circuit for, 173
pulse width modulator, 171

Q
quad D-type flip flop, 201

R
R-S flip flop, 14, 15
random number generator, 132
random tone generator, 107
rectangular waves, 172
resistance
   programmable, 54
   variable, 53
resistor-transistor logic (RTL), 20
revolution counter, 140
roulette wheel, 162

S
Schmitt trigger, 67, 69
SCR, 52
selectable delay, 167
seven-stage ripple-carry binary counter, 194
shift registers, 28-33
signal generator and music-making projects, 84-116
simple frequency meter, 63
sine wave generator, 92
sine waves, digital approximation of, 93
single digit decimal counter, 118
SIPO shift register, 31
SISO shift registers, 28, 29
SPDT switch, 41
SPST switch, 40
square waves, 88
stepped waves, 88
stepped-wave generator, 88
strings, 2
substituting components, viii
switch debouncer, 145, 168
   instructions for, 166-167

T
ten-step tone sequencer, 103
test equipment, 59-75
   logic probe as, 59-62
   simple frequency meter, 63
theremin, 96
   photo-, 98
Theremin, Leon, 96
time keeping circuits, 149-158
timebase ICs, 150
timebase signals, 68
timed touch switch, 43

touch switches, 42-53
    appliance controller, 52
    automatic night light, 50
    built-in delay for, 45
    counter using, 141
    light activated gate for, 46
    pulse delayer as, 47
    timed, 43
touch-switch counter, 141
toy organ, 96
transistor-transistor logic (TTL), 20
triangle waves, 88

triangle-wave generator, 90
triggering, 13
triple three-input NAND gate, 193
truth table, 12
tunable oscillator, 86

U

up/down counter, 202

V

variable resistance/capacitance, 53
    programmable, 53

VOM, 59

W

waveforms
    rectangular, 172
    sine, 93
    stepped and square, 88
    triangular, 88, 90
window counting method, 67

X

XOR gates, 8, 9
    digital comparator use of, 24

Edited by David M. Gauthier

Parts List

14583 Schmitt trigger, 71
1N4148, 58, 139
1N4734, 75
1N914, 58

2N2222, 58, 67, 86, 99, 139, 164
2N3302, 71
2N3904, 58, 86, 99, 139
2N3906, 95
2N5826, 71

555 timer, 133, 161
556 dual timer, 71, 109, 144

74C123, 139
74C143, 139
74C154, 82, 163, 164
74C174, 144
74C175, 30, 32, 163, 164, 201
74C193, 80, 82, 131, 202
74C41, 67
74C46, 131

74C74, 139
74C85 digital comparator, 27
74C90, 75, 139
74C92, 139
7555 timer, 48, 63-67, 75, 103, 105, 109,
  133, 134, 147, 148, 161, 163, 170, 173, 176, 203
CD4001, 78, 81, 103, 161, 168, 176, 182, 188
CD4009, 131, 164
CD4011, 35, 44, 46, 65, 67, 71, 75, 81, 86,
  89, 97, 99, 103, 144, 147, 148, 154, 157, 170, 189
CD4013, 17, 103, 175, 176, 190
CD4017, 48, 92, 105, 124-127, 152, 157, 160, 161, 170, 182, 191
CD4018, 92, 93, 94, 95, 192
CD4023, 193
CD4024, 194

CD4026, 71
CD4027, 195
CD4028, 196
CD4029, 197
CD4040, 115, 198
CD4046, 86, 87
CD4049, 35, 51, 52, 58, 91, 144, 157, 161, 183, 199
CD4051, 115
CD4066, 54, 103, 105, 200
CD4070, 178
CD4081, 153
CD40ll, 45
CD4511, 75, 119, 120, 133, 134, 144, 147, 157
CD4514, 109
CD4518, 119, 120, 133, 134, 144, 147, 157
CD4528, 89

MM5369, 139, 150, 152, 153

210

Other Bestsellers For 2995

☐ **EXPERIMENTS IN ARTIFICIAL NEURAL NETWORKS**—Ed Rietman

Build your own neural networking breadboards—systems that can store and retrieve information like the brain! This book shows you how to use threshold logic circuits and computer software programs to simulate the neural systems of the brain in information processing. The author describes artificial electronic neural networks and provides detailed schematics for the construction of six neural network circuits. The circuits are stand-alone and PC-interfaced units. 160 pp., 80 illus.
Paper $19.95　　　　　　　　　　　　　　Hard $24.95
Book No. 3037

☐ **101 SOLDERLESS BREADBOARDING PROJECTS**—Delton T. Horn

Would you like to build your own electronic circuits but can't find projects that allow for creative experimentation? Want to do more than just duplicate someone else's ideas? In anticipation of your needs, Delton T. Horn has put together the ideal project *ideas* book! It gives you the option of customizing each project. With over 100 circuits and circuit variations, you can design and build practical, useful devices from scratch! 220 pp., 273 illus.
Paper $18.95　　　　　　　　　　　　　　Hard $24.95
Book No. 2985

☐ **50 POWERFUL PRINTED CIRCUIT BOARDS PROJECTS**—Dave Prochnow

If you've ever experienced the frustration and disappointment of failed projects or are interested in finding practical, unique electronic devices, then you won't want to miss this book! Here, in a single reference, are 50 fully described, and detailed electronics projects, complete with schematic diagrams and instructions. More importantly, *50 Powerful Printed Circuit Board Projects* provides what few others on the market can. *Each project in this book comes with its own computer generated photo image of the printed circuit board!* 208 pp., 184 illus.
Paper $18.95　　　　　　　　　　　　　　Hard $23.95
Book No. 2972

☐ **THE DIGITAL IC HANDBOOK**—Michael S. Morley

This book will make it easier for you to determine which digital ICs are currently available, how they work, and in what instances they will function most effectively. The author examines ICs from many major manufacturers and compares them not only by technology and key specification but by package and price. And, if you've ever been overwhelmed by the number of choices, this book will help you sort through the hundreds of circuits and evaluate your options—ensuring that you choose the right digital IC for your specific needs. 624 pp., 273 illus.
Paper $34.50　　　　　　　　　　　　　　Hard $49.50
Book No. 3002

☐ **HOW TO DESIGN SOLID-STATE CIRCUITS**—2nd Edition—Mannie Horowitz and Delton T. Horn

Design and build useful electronic circuits from scratch! The authors provide the exact data you need on every aspect of semiconductor design . . . performance characteristics . . . applications potential . . . operating reliability . . . and more! Four major categories of semiconductors are examined: Diodes . . . Transistors . . . Integrated Circuits . . . Thyristors. This second edition is filled with procedures, advice, techniques, and background information. All the hands-on direction you need to understand and use semiconductors in all kinds of electronic devices is provided. Ranging from simple temperature-sensitive resistors to integrated circuit units composed of multiple microcircuits, this new edition describes a host of the latest in solid-state devices. 380 pp., 297 illus.
Paper $19.95　　　　　　　　　　　　　　Hard $24.95
Book No. 2975

☐ **EXPERIMENTS WITH EPROMS**—Dave Prochnow

One of the greatest versatilities in advanced circuit design is EPROM (Erasable-Programmable Read-Only-Memory) programming. Now, Dave Prochnow takes an in-depth look at these special integrated circuits (ICs) that can be user-programmed to perform specific applications in a microcomputer. Fifteen fascinating experiments are a special feature of this book that presents not only the technology but also explains the use of EPROMs. 208 pp., 241 illus.
Paper $19.95　　　　　　　　　　　　　　Hard $24.95
Book No. 2962

Other Bestsellers From TAB

☐ **44 POWER SUPPLIES FOR YOUR ELECTRONIC PROJECTS**—Robert J. Traister and Jonathan L. Mayo

Here's a sourcebook that will make an invaluable addition to your electronics bookshelf whether you're a beginning hobbyist looking for a practical introduction to power supply technology, with specific applications . . . or a technician in need of a quick reference to power supply circuitry. You'll find guidance in building 44 supply circuits as well as how to use breadboards, boards, or even printed circuits of your own design. 220 pp., 208 illus.
Paper $18.95   Hard $24.95
Book No. 2922

☐ **500 ELECTRONIC IC CIRCUITS WITH PRACTICAL APPLICATIONS**—James A. Whitson

More than just an electronics book that provides circuit schematics or step-by-step projects, this complete sourcebook provides both a wealth of practical electronics circuits AND the additional information you need about specific components. You will be able to use this guide to learn basic project-building skills and become familiar with some of the popular ICs. 336 pp., 600 illus.
Paper $22.95   Hard $20.95
Book No. 2920

☐ **BEYOND THE TRANSISTOR: 133 ELECTRONICS PROJECTS**—Rufus P. Turner and Brinton L. Rutherford

Strongly emphasized in this 2nd edition are the essential basics of electronics theory and practice. This is a guide that will give its reader the unique advantage of being able to keep up to date with the many rapid advances continuously taking place in the electronics field. It is an excellent reference for the beginner, student, or hobbyist. 240 pp., 173 illus.
Paper $12.95   Hard $16.95
Book No. 2887

☐ **OSCILLATORS SIMPLIFIED, with 61 Projects**—Delton T. Horn

Here's thorough coverage of oscillator signal generator circuits with numerous practical application projects. Pulling together information previously available only in bits and pieces from a variety of resources, Horn has organized this book according to the active devices around which the circuits are built. You'll also find extremely useful information on dedicated oscillator integrated circuits (ICs), an in-depth look at digital waveform synthesis, a clear description of the phase locked loop (PLL), and plenty of practical tips on troubleshooting signal generator circuits. 238 pp., 180 illus.
Paper $14.95   Hard $17.95
Book No. 2875

*Prices subject to change without notice.

Look for these and other TAB books at your local bookstore.

TAB BOOKS Inc.
Blue Ridge Summit, PA 17294-0850

Send $1 for the new TAB Catalog describing over 1300 titles currently in print and receive a coupon worth $1 off on your next purchase from TAB.

OR CALL TOLL-FREE TODAY: 1-800-233-1128
IN PENNSYLVANIA AND ALASKA, CALL: 717-794-2191